Liquid Metal Alloys in Electronics

David J. Fisher

Copyright © 2020 by the author

Published by **Materials Research Forum LLC**
Millersville, PA 17551, USA

Published as part of the book series
Materials Research Foundations
Volume 70 (2020)
ISSN 2471-8890 (Print)
ISSN 2471-8904 (Online)

Print ISBN 978-1-64490-068-0
ePDF ISBN 978-1-64490-069-7

Distributed worldwide by

Materials Research Forum LLC
105 Springdale Lane
Millersville, PA 17551
USA
http://www.mrforum.com

Printed in the United States of America
10 9 8 7 6 5 4 3 2 1

Table of Contents

Introduction 1

Galinstan 9

EGaIn 21

Self-Assembled Monolayers 35

Energy-Harvesting 48

Reconfigurable Antennae..51

Patch Antennae 61

Sensors..83

Conformable Electronics...93

References 108

Keyword Index *130*

About the author..*133*

Table of Contents

Introduction

Metals which are liquid at room-temperature form a class of materials which is of rapidly increasing interest. This is because such materials combine the high electrical conductivity of metals with the ease of manipulation and reconfiguration of liquids.

For centuries, both roles have been fulfilled by mercury; one immediately thinks for example of Faraday's disc motor, not to mention mercury switches and other useful electrical devices. Mercury has fallen out of favour however, because of its toxicity, especially in applications where it may escape, in either liquid or vapour form, from its containment.

The present work discusses the applications of its replacements, which consist largely of gallium-based alloys. When it comes to the task of finding a replacement for mercury, it has been found that only two alloys fit the bill. Among the pure metals that are liquid when near to room temperature, two of them (cesium, rubidium) are highly reactive and another (francium) is radioactive, thus leaving only gallium ... and preferably its alloys. Strictly speaking, gallium should not be on the shortlist, given its solidity at standard temperature and pressure. It has a marked tendency to undercool, by close to the theoretical maximum value (about 20% of the absolute melting-point), however and may often be unexpectedly found in the liquid state.

It will be found that the use of gallium alloys is not without problems because, although the gallium alloy may be liquid, a skin of solid oxide can form at the surface. This oxide can interfere with electrochemical measurements, alter the properties of the surface, and affect the fluid dynamics in a way that had once been thought to be a considerable problem. Perhaps it was the likelihood of problems caused by oxide residues which caused Daimler AG to discontinue its attempt to patent the concept of 'a pane for a motor vehicle with a hollow chamber that stores fluid for the darkening pane where the fluid consists of Galinstan'[1]. On the other hand, the oxide confers some useful properties. It does, for example, provide sufficient rigidity to counterbalance the effect of surface tension and prevents this liquid metal from forming spherical beads in the same manner as does mercury. The solid oxide skin even allows the liquid to be moulded to some extent, and permits the liquid to be used in novel roles as soft electrodes, shape-reconfigurable conductors, stretchable antennae, sensors, wires, interconnects and self-healing circuits. The oxide layer, which forms almost instantly in air under ambient conditions, is at least 0.7nm thick and does not thicken appreciably with time in dry air. The Pilling-Bedworth ratio of the gallium alloys does not appear to be known but, overall, the rapidity of oxidation and the stability of the oxide recalls the case of aluminium: a metal which would of course constantly fizzle away like sodium if it were not for the fortuitous characteristics of its oxide.

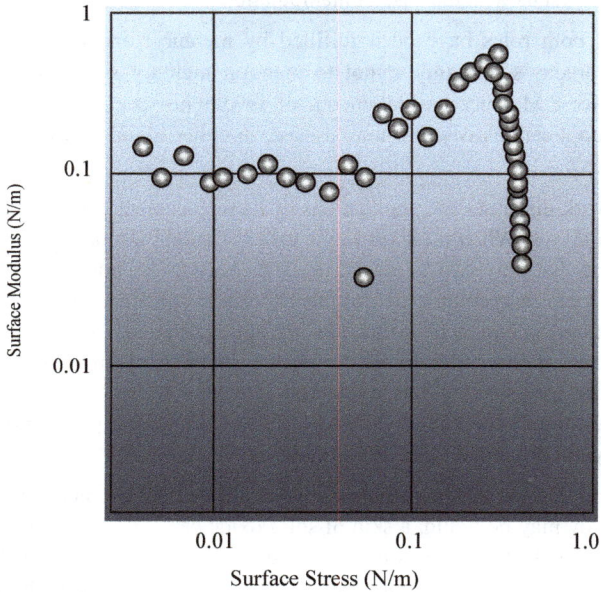

Figure 1. Surface elastic modulus versus surface
stress plot for a typical gallium-indium alloy

The resemblance of gallium to aluminium, and many of its other characteristics, was recognized long ago[2]:

A most remarkable property of gallium is its low melting point; it liquefies completely at 86°Fah, or below the heat of the hand, and, still more curiously, when once melted at this temperature, it may be cooled down even to the freezing point of water without solidifying, and be kept unchanged in the liquid state for months. Indeed, in the original communication of its discovery to the French Academy, it was described as a new liquid metal, similar to mercury; but on touching with a fragment of solid gallium a portion of the liquid metal in this state of so-called surfusion[1], it at once solidifies. Unlike lead again, gallium is a highly crystalline metal, its form being that of a square octahedron. In its chemical habitudes the rare element gallium shows the greatest analogy to the abundant element aluminum. In particular it forms a sort of alum not to be distinguished in its appearance from ordinary alum, but containing oxid [sic] of gallium instead of oxid [sic] of aluminum or alumina. Mendelejeff in 1871 ... predicted, for example, that the specific gravity of the missing metal would prove to be about 5.9. Operating on very small quantities, M. De Boisbaudran, in the first instance, found the specific gravity of gallium to be 4.7; but on repeating his determination in 1876, with special precautions and on a somewhat larger though still very small scale, he found it to be exactly 5.936 – certainly a most remarkable fulfillment of the prediction.

On the other hand, the same issue of this magazine strangely asserted elsewhere (p159) that gallium was harder than iron. It was also noted that gallium alloys had been used in automatic fire alarms and that it was planned to use gallium itself in thermometers operating between 100 and 1000F.

Although the critical factor determining the mechanical properties of EGaIn is the rapid formation of a passivating oxide layer upon exposure to ambient air, the surface reactivity of EGaIn with oxygen and water-vapour at the molecular level is little explored. Ambient-pressure X-ray photo-electron spectroscopy of the liquid/gas interface of EGaIn in oxygen and water vapour has shown[3] that the oxygen and water vapour both react with the gallium of EGaIn so as to form Ga_2O_3 as an outer layer and Ga_2O as an interlayer. Although similar products are formed, there are marked differences in the pressure-dependence and adsorbate-induced binding-energy shifts.

Because the bulk viscosity of a gallium-indium alloy is typically low, the properties of its oxide trump its own mechanical behaviour. When it is held between the oscillating surfaces of a parallel-plate rheometer therefore, increasing the amplitude of the oscillations without changing their frequency permits the deduction of the stress-strain relationship of the oxide. Plotting the magnitude of the surface elastic modulus as a function of the surface stress (figure 1) then reveals that there exists a critical yield surface stress, below which the oxide layer is elastic. When this critical yield stress is exceeded, the oxide is broken up and the metal flows easily[4].

Suitable control of the environment can prevent the formation of the oxide skin, or remove it. This then permits the selective wetting of gallium-lyophilic surfaces; an aim

---

[1] Undercooling

which can also be achieved by altering the wettability of the liquid metal by changing the chemistry and topology of a substrate so as to make it superlyophobic or superlyophilic. This topic is of interest in its own right, in that experience with water has shown that an extremely low wettability leads, for example, to a self-cleaning capability. Inspired by this, much effort has been devoted to the creation of non-wetting super-hydrophobic or super-lyophobic surfaces by giving them an hierarchical micro/nano dual-scale structure.

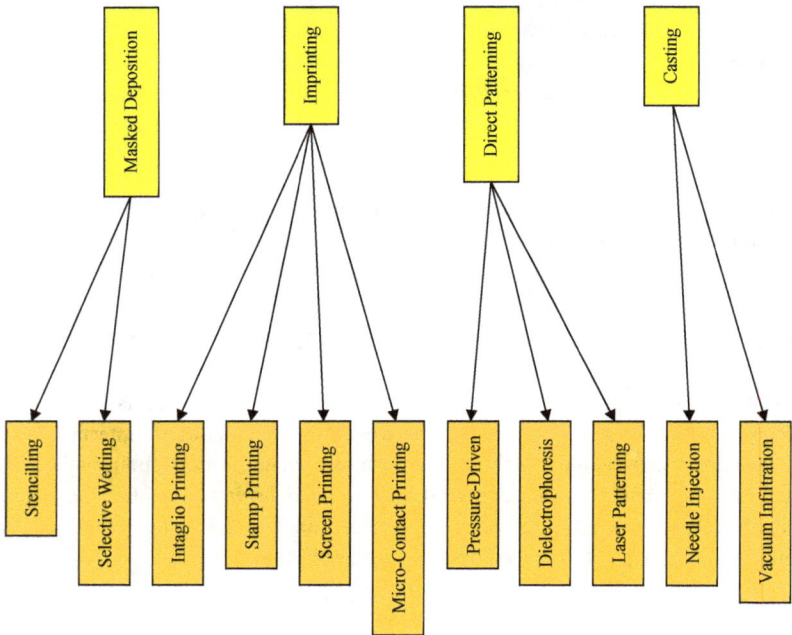

Figure 2. Reported methods for embedding liquid metal in an elastomer

The selective wetting deposition technique (figure 2) uses a sacrificial layer to wet a substrate with liquid metal selectively. A polytetrafluoroethylene particle-filled photoresist, tin foil and an elastomer are layered on a glass slide before patterning the polytetrafluoroethylene and tin foil photolithographically. When liquid metal is applied to the suitably prepared tin foil, it wets it reactively but does not adhere to the photoresist.

This technique tends to produce rough edges and offers only a limited choice of geometries. In this connection, the wetting behavior of droplets of gallium-indium on thin metal films, especially tin and indium, was studied[5] and this showed that the droplets reactively wetted the thin metal foils. The wettability could however be varied by changing the surface texture. The effects of the composition and texture of the substrate upon wetting were deduced by measuring the contact angle and contact diameter as a function of time.

Here again unfortunately the comparative mechanical stiffness of the oxide complicates the use of traditional methods, such as contact-angle and sessile drop, to measure interfacial properties.

Running quickly now through the other preparation methods: imprint processing is a simple method for producing patterns of liquid metals and, as the figure shows, can be sub-divided into various variants.

Recess printing, here termed intaglio printing, involves applying a thin layer of the liquid metal to a flat surface. A polydimethylsiloxane mould bearing the desired features is meanwhile produced. The latter mould is then pressed against the liquid layer and, due to the tenacious nature of the gallium alloy, the recesses become filled. A sealant layer can then ideally be applied in order to encapsulate the patterned structure. The liquid metal unfortunately has a tendency to wet the non-recessed parts of the polydimethylsiloxane mould, and one problem with the method is that residues of the liquid can be present, away from the channel pattern. This is thus a situation which would benefit from the use of one of the above-mentioned techniques for imparting super-hydrophobicity to the mould surface.

Stamp printing is very much like the above technique, but instead involves using the raised pattern of a stamp to transfer the liquid metal to another surface. A polydimethylsiloxane stamp which bears the desired pattern is first prepared by casting. A simple paint-brush is then used to 'ink' the raised features of the stamp with liquid metal. The stamp is pressed onto the substrate and the liquid-metal pattern is thus transferred. At this point it may be advantageous to freeze the liquid, lest subsequent processing should disturb the pattern, bearing in mind that some common alloys can be frozen in a refrigerator while others require a freezer. A border is then placed around the patterned structure and spin-coating and curing techniques are used to apply an elastomer sealant layer over the patterned, and possibly frozen, alloy. Another potential problem can arise due to uneven wetting of the stamp features, or of the surface to which the pattern is to be transferred. This can lead to gaps in the pattern or to excessively thick sections.

Screen printing is essentially the same traditional silk-screen technique that is used to decorate T-shirts. In one version, the screen is prepared via the potassium hydroxide etching of silicon to produce micro-scale holes. When a pressurized reservoir of the liquid metal is placed against the screen, the liquid metal bulges through the 40 to 70μm holes in the screen and is deposited onto the target surface; a surface which has been pre-patterned with gold in order to encourage wetting by the liquid metal. When the screen is removed, a pad forces the metal to form adherent droplets.

Micro-contact printing involves the use of a hemispherical polydimethylsiloxane tip as a delivery nozzle, mounted on a print-head which can be moved in 3 dimensions under computer-control. The tip is then repeatedly dipped into the liquid metal, collecting an approximately 300μm bead of liquid each time, and used to transfer the droplet to the required point on the target substrate. The technique obviously depends critically upon the adhesion of the liquid metal to a polydimethylsiloxane substrate, and the droplets coalesce into a continuous line. Once solidified, the pattern can be covered with a layer of polydimethylsiloxane. The main disadvantage of this method is the repetitive, rather than continuous, printing process. The most recent progress[6] involves a microcontact printing process for the reproducible manufacture of liquid-alloy based stretchable electronics. The process uses surface-functionalized reusable stamps, rigid or deformable, to transfer eutectic gallium–indium patterns onto elastomer substrates selectively without leaving residues or creating electrical short-circuits. In this way, EGaIn features which are as small as 15μm in line-width can be produced.

Direct patterning methods can also be sub-divided into 3 types. In the pressure-driven technique, simple pens were originally used to write low-resolution designs in metal ink directly onto various substrates. High-resolution designs were later drawn by using a computer-controlled roller-ball printer. Pressure exerted on the liquid metal and roller-ball lead to uniform transfer of the liquid to the substrate, where adhesion to a flexible polyvinyl chloride substrate is aided by oxidation of the metal. The pattern can then be transferred to polydimethylsiloxane film and sealed using another layer of polydimethylsiloxane. One alternative direct-printing method is to spray the liquid metal onto a substrate. Another is to use a syringe pump to feed the liquid through a nozzle. Whatever the details of the method used, the smallest feature size obtained is of the order of 45μm. The main disadvantage is the relative complexity of the apparatus.

The phenomenon of dielectrophoresis, best known in connection with biological-cell separation, can be used to create 3-dimensional structures from liquid metal by first producing 0.1 to 0.35μm diameter micro-droplets of a suitable gallium-indium alloy in de-ionized water by means of ultra-sonication. The suspension of micro-droplets is then placed on a dielectrophoresis platform equipped with pairs of 80μm-diameter planar chromium/gold micro-electrode pads, having a separation of 40μm, with a similar micro-electrode island between each pairs of pads. The electrically conducting droplets can typically be immobilized between the pads when a sinusoidal signal with an amplitude of 15V and a frequency of 20MHz is applied. The immobilization is due mainly to the dielectrophoresis force. Sodium hydroxide solution is then used to remove some of the oxide which forms on the micro-droplets, thus allowing them to amalgamate with the gold micro-electrodes. A high rate of flow is then used to wash away the unmerged droplets, resulting in the formation of 3-dimensional liquid-metal micro-electrodes with a spherical end-cap. The mechanical properties of biological cells are closely related to their physiological function, and abnormal properties can thus be an indicator of abnormal cellular behaviour. A device was constructed, for measuring cell mechanical properties, by using dielectrophoretic force[7]. The electrodes which are conventionally used for stretching cells are made from indium tin oxide, gold or platinum, but a superior galinstan microelectrode which offered the advantages of easy manufacture, low cost, stability under high voltages and reusability, could be integrated into a microfluidic chip and a modulated sine-wave voltage could be applied to the chip so as to stretch red blood-cells. Physiological activity could then be deduced by measuring the cell mechanical properties.

Liquid-metal electrodes can be directly patterned by using a commercial CO_2 laser cutter to remove the metal subtractively. A thin layer of the liquid, is encapsulated between 2 layers of polydimethylsiloxane; the upper layer serving to aid material removal by strongly absorbing the laser radiation. This layer also largely prevents oxidation and other contamination during cutting. The energy of the laser beam vaporises both polydimethylsiloxane layers, with the vaporized elastomer escaping to leave a pattern of liquid metal features having features which can be as fine as 100μm. The latter are held in place by the native Ga_2O_3 film until a new layer of polydimethylsiloxane can be cast over the structure to give completely encased electrodes.

Casting in the present context is essentially the same as the traditional macroscopic process. That is, the liquid metal is introduced by some means into a mould containing a suitably shaped cavity and allowed to solidify. Needle injection permits much greater control of pattern dimensions than do any of the above methods, but this is off-set by the complexity of producing a suitable mould; particularly with regard to ensuring interconnection of all of the parts of the mould. The latter is essential given that complete filling has to be achieved using a single injection. In one version of the process, a flat glass plate is pressed against a polydimethylsiloxane mould having a form which has been produced by using laser ablation or some other method. A syringe is then inserted at each end; one to inject the liquid and the other to help air to escape. The trapping of air in complex structures nevertheless remains a common problem. The frozen pattern of metal can then simply be embedded in an elastomer. On the other hand, it is also possible to imitate a technique which is routine in the field of lost-wax casting. It is possible there to take standard injection-mould wax shapes and melt them together into any desired form before investing them with a ceramic slurry. In the present case, various independent patterns can be deformed and joined as desired before coating the whole assembly with polydimethylsiloxane. Although somewhat labour-intensive, and requiring that the work be performed in a low-temperature low-humidity environment, this process can produce shapes of good dimensional accuracy which are distributed throughout an entire block of elastomer. The final encapsulation step may also cause problems, given that the curing of the polydimethylsiloxane can warm the assembly to a considerable degree, leading to the possible distortion of delicate features.

The vacuum-assisted infiltration of liquid metal, which is best for the creation of fine 3-dimensional deep features, is described later. It completely avoids the problem of air-trapping which occurs in the case of needle injection.

Table 1. Properties of gallium-indium alloys and mercury

Liquid	MP (C)	Viscosity (mm^2/s)	Surface Tension (mN/m)	Conductivity (S/m)
EGaIn	15.5	0.32	624	0.34×10^7
Galinstan	-19	0.37	535	0.38×10^7
Mercury	−39	0.11	428	0.10×10^7

Much of the interest in liquid-metal devices arises from their potential usefulness in the field of so-called soft and stretchable electronics. Developments in this rapidly growing field are motivated by the possibilities which are offered by devices that have mechanical properties which are similar to those of living tissue and textiles. Liquid metals are essentially infinitely deformable but retain their metallic conductivity. They can therefore be used for manufacturing stretchable wires and interconnects, conformable electrodes, reconfigurable antennae and self-healing circuits; examples of most of which are discussed later. These can also be potentially used for the construction of components which consist entirely of so-called soft materials.

That this field has been expanding rapidly can be seen from the fact that, as recently as 7 years ago, it was still being declared that, '*conductive electrodes and electric circuits that can remain active and electrically stable under large mechanical deformations are highly desirable for applications such as flexible displays, field-effect transistors, energy-related devices, smart clothing and actuators. However, high conductivity and stretchability seem to be mutually exclusive parameters*'[8]. The most promising solution at that time was to use 1-dimensional materials such as carbon nanotubes and metal nanowires on a stretchable fabric, wavy metal strips, elastomers loaded with conductive fillers and interpenetrating networks of liquid metal and rubber. The conductivities at large strains remained too low for practical application. One particular solution was a composite mat of silver nanoparticles plus rubber fibres which was compatible with any substrate and was suitable for large-area applications. Silver nanoparticles were here incorporated into electro-spun polystyrene-block butadiene-block styrene rubber fibres and were then directly converted *in situ* into silver nanoparticles within the fibre mat. Percolation of the silver nanoparticles within the fibres then led to a high (2200S/cm) bulk conductivity, in a 150µm-thick mat, which was retained at 100% strain. The current materials, based upon liquid metals that are incorporated into elastomers by using microfluidic methods, are clearly superior. The growing interest in deformable electronics has driven rapid advances in the development of stretchable conducting composites for use in electric circuits, interconnects and electrodes.

Almost in parallel with the development of room-temperature liquid metals there has indeed occurred the development of microfluidics, another growing field of technology which aims to control fluids at the micro-scale. When combined with liquid-metal technology, the prospects for applications in deformable electrodes, wearable electronics and flexible antennae are excellent. Most of the current barriers to universal exploitation revolve around the problems which are associated with surface oxidation of the alloys.

It transpires, as noted above, that there are really only two contenders for use as room-temperature liquid alloys and they are introduced below.

Galinstan

Given that this is one of the most popular liquid metal alloys, together with EGaIn, in the present context, it will be useful to consider it in some detail, particularly as the uses to which it is put constitute a handy summary of the applications of all liquid metal alloys.

The proprietary alloy, Galinstan, is an eutectic alloy of gallium, indium and tin (Ga-21.5In-10%Sn) which melts at -19C and has several advantages over mercury (table 1), such as non-toxicity (with a very low vapour pressure), higher thermal conductivity and lower electrical resistivity. Its thermal conductivity is some 27 times that of water, its dynamic viscosity is only twice that of water and its volumetric specific heat is about half of that of water. Its surface tension is some 10 times that of water. The initial attraction of the alloy was therefore its good heat-transfer coefficient, although this is offset by its tendency to oxidize and by its low specific heat. It is therefore preferred to disperse nanoscale galinstan droplets in a fluid for use as a coolant, and the fluidity of the liquid metal particles largely avoids the usual colloidal-fluid problems of rapid sedimentation, erosion and clogging. Ultrasonic emulsification can produce nanodroplets which are of the order of hundreds of nanometres in size, or less. Extended exposure of galinstan to the oxygen in water unfortunately resurrects the problem of oxidation, which can be severe in the case of droplets. When metallic nano-emulsions are considered as microchannel heat-sink working fluids, the effect of the high thermal conductivity of galinstan is evident. Convective heat-transfer is greatly improved, while the high viscosity due to nanoscale blending and the inherently low specific heat of galinstan reduces the flow-rate and thereby increases the caloric thermal resistance.

Although many applications are based upon the conformable behavior of liquid metals such as galinstan, relatively little work has been performed on understanding its solidification with a view to characterizing its true operating range. Recent work[9] has explored the phase-change of bulk galinstan by means of rheological methods, X-ray scattering and differential scanning calorimetry. This revealed that there exists a gap of 20C between the solidification of galinstan, at about -10C, and its melting at 10C. Crystallization is observed below -10C, suggesting an ordering of the gallium, indium and tin phases which is unstable under repeated solidification-melting cycles and perhaps a structural rearrangement below -30C. Dispersions of galinstan in polydimethylsiloxane remain soft at below -20C, whereas galinstan channels in polydimethylsiloxane appear to solidify at length-scales as small as 500μm. The phase-change behaviour was explored between -20C and 50C; with holding at -20C for 0.5h. All of the test-channels had the same length but various cross-sections and depths. The largest (1mm x 2.5mm) channel exhibited freezing at -8.3C, and the beginning of melting at 6.9C. The intermediate (1mm x 1mm) channel exhibited freezing initiation at -15.4C and the onset of melting at 9.0C. These freezing and melting points corresponded to those deduced from bulk rheology data. As in the case of all such measurements, it is possible that the figures were affected by heat transfer around the low thermal-conductivity channel and by the availability of nucleation sites. The general behaviour of galinstan in channels which were larger than 1mm^2 nevertheless resembled that of bulk galinstan. The smallest (0.5mm x 0.5mm) channels exhibited a very slight increase in modulus at -20C after being held for more than 1h. The change in solidification behaviour as a function of channel-size suggested that, in small volumes of galinstan, the material remained soft even after solidification or that galinstan solidification is kinetically driven at small length-scales. This also suggested that, if dispersions were held at -20C for perhaps tens of hours, solidification would eventually occur. The overall conclusion is that galinstan can exhibit a liquid-like behaviour down to -10C while, at lower temperatures, its solidification occurs at temperatures which depend upon its environment. When solidification does occur, galinstan has to be reheated by a substantial amount in order to restore its liquid-like behaviour.

Heat-dissipation is a key aim in producing reliable high power-density electronic systems. Devices which are capable of managing heat transfer are required in order to ensure optimum heat-dissipation and reliability by isothermalization. An integrated liquid cooling system can be created[10] by placing a small droplet of galinstan over a hot-spot in an electronic device, for example. Energizing the droplet using a square-wave signal then creates a surface-tension gradient across the droplet and in turn induces Marangoni flow over the droplet surface. This produces a high flow-rate of coolant through the cooling channel and a so-called soft-pump effect. The high thermal conductivity of the liquid metal also extends the heat-transfer surface and aids heat dissipation. The overall effect is to provide high flow-rates with low power consumption. A mm-scale liquid-metal droplet thermal switch was developed[11] which could control heat transfer, and this was integrated with a single gallium nitride device, having a 2.6mm x 4.6mm area on a printed-circuit board, which dissipated 1.8W. The thermal switch could control the heat transfer by conducting 1.3W in the ON mode when the gallium nitride was at 51C, and conducting 0.5W in the OFF mode when the device was at 95C. A 1-dimensional thermal-resistance model, combined with a 3-dimensional finite-element simulation, gave excellent agreement with experimental data. When the thermal switch was integrated with two gallium nitride devices, it could balance the heat-transfer rate and improve junction-temperature uniformity and system reliability by reducing the inter-device temperature difference from more than 10C, without the switch, to just 0C.

The performance of a hybrid solar system can be improved by using galinstan as a coolant. Simulations and experiments were used[12] to measure criteria, such as electrical efficiency, thermal efficiency and exergy efficiency for galinstan, water and air coolants, as a function of ambient temperature, fluid inlet velocity and solar radiation. The exergy, thermal and electrical efficiencies were at least 11, 12 and 12% higher, respectively, for a hybrid system which used galinstan as a coolant, as compared with air or water coolants.

A galinstan-based minichannel cooling technique has recently been developed[13] for high heat-flux thermal management. A compact electromagnetic pump with no moving parts provides a pressure head of up to 100kPa to a minichannel heat sink having a width of

1mm and a height of 5mm. The system can dissipate heat with a heat flux of $300W/cm^2$ and a heat power of 1500W. As compared with water-based minichannel cooling, the galinstan-based system provides greater heat-transfer, and the pressure loss in a galinstan-based minichannel heat sink is much lower than that for water-based microchannel cooling. The heat capacity thermal resistance also becomes an important factor, in the thermal performance of galinstan-based minichannels, due to the high thermal conductivity and lower heat capacity of galinstan.

Microchannel heat-sinks are relevant to thermal management because the combined effect of surface-area increase and small length-scales results in low wall-to-bulk temperature differences. A thermal resistance of 0.09C/W was possible when a heat-flux of $790W/cm^2$ was imposed on a 1cm x 1cm footprint of a 400μm-thick silicon substrate by using single-phase water-based microchannel cooling and a 214kPa pressure difference to drive the flow. In the case of galinstan-based heat sinks, the optimum channel-widths are of the order of hundreds of micrometers, rather than tens, and the above thermal resistance of 0.09C could be equalled. An extremely low thermal resistance and high heat-flux were achieved by using single-phase water-based microchannel cooling: 0.071C/W and $1003W/cm^2$, respectively. Early experimental data[14] on galinstan-based minichannel heat sinks indicated a thermal resistance as low as 0.077C/W at a heat-flux of $1214W/cm^2$, with a maximum heat flux of $1504W/cm^2$.

An analytical study was made[15] of the effect of using structured surfaces to reduce the overall thermal resistance of galinstan-based micro-gap cooling in the laminar flow regime. The high surface tension of galinstan, which is 7 times that of water, implies that it can be stable in a non-wetting state at the pressure differences required for ensuring flow through microgaps. The flow over the structured surfaces encountered a limited liquid/solid contact area and a low-viscosity gas layer between the channel walls and the galinstan. Consequent reductions in the friction factor resulted in a decreased caloric resistance, but an associated reduction in the Nusselt number increased the convective resistance. A further numerical study was made[16] of the laminar flow and forced convective heat-transfer of galinstan through a mini-channel heat-sink which was exposed to a constant heat-flux. This involved detailed parametric analysis of the effect of the heat-sink geometry and of the inlet velocity upon the pressure-drop, pumping power and maximum heat-flux and led to prediction of the optimum heat-sink dimensions and inlet velocity. Comparison of the numerical results with analytical correlations revealed points of both agreement and disagreement. The flow properties and heat-transfer performances of mini-channel heat-sink cooling, using liquid metal, nanofluid and water were compared; again demonstrating the advantages of liquid-metal coolant.

A miniature apparatus has been used[17] to study the characteristics of liquid-metal free surface flow and laminar convective heat transfer on various surface structures. This showed that, within a certain velocity-range, galinstan could flow in the form of an even liquid film on acid-pickled stainless-steel surfaces. As the flow-rate increased, the convective heat transfer coefficient increased.

Another reason for the popularity of galinstan is the possibility of replacing mercury, particularly when the toxicity of the latter is a problem. Here again, the tendency to oxidation of galinstan limits its wider use. Unlike mercury, which wets few surfaces, oxidized galinstan wets almost everything. In particular, its tendency to wet human skin can make it unpleasant to handle even though it is non-toxic. In order to avoid the wetting problem when used in electronics as a coolant, micro-pillar array-based super-lyophobic polydimethylsiloxane micro-tunnels were proposed[18] for microfluidic use. The aim of microfluidics is to manipulate fluids at the sub-mm scale. The formation of stable fluid metal structures in microchannels permits the creation of various components, such as antennae. Electrodes in microchannels are useful for applying electric fields. Liquid metals can be used to make metallic components which are in direct contact with microchannels. Alignment with microchannels is difficult because of the small length-scales which are involved. The ability to inject the metal into channels makes it possible to form electrodes in a single step, without requiring alignment. Electrodes can be brought into direct contact with the fluid by using posts to separate the metal from the fluid: the oxide spans the posts, prevents the metal from flowing into the main channel, and mechanically stabilizes the electrode.

Various pitch distances and micro-pillar arrays were used to evaluate the lyophobicity of galinstan in terms of contact-angle and sliding-angle. Contact-angle analysis revealed the extent of reduction of the wetting behaviour of the oxide. The highest advancing-angle, 155°, and receding angle, 136°, were found for 200μm-thick polydimethylsiloxane film and 37wt%HCl solution. The galinstan in a microfluidic channel was surrounded by a coplanar channel filled with HCl solution. Due to the permeability of polydimethylsiloxane, the HCl permeated the wall between the two channels and permitted continuous chemical reaction with the oxidized galinstan[19].

Various reactions can be involved here:

$$HCl + Ga_2O + Ga_2O_3 \Rightarrow GaCl_2 + GaCl_3 + H_2O$$

$$HCl + In_2O_3 \Rightarrow InCl_3 + H_2O$$

$$HCl + SnO \Rightarrow SnCl_2 + H_2O$$

The most important reaction is that of the gallium oxides to give gallium chlorides, because gallium is the principal oxide which is formed on either galinstan or EGaIn and the chlorides do not exert the same effects on flow and wettability and this permits a drop to behave in a manner closer to that of a pure liquid. The HCl can be used in the form of a solution, or vapour. It was noted some time ago[20] that the metallic residues which were left behind on surfaces that had come into contact with gallium-based liquid metal alloys might limit the applications of the alloys. One method which was proposed for the control of the wetting characteristics of gallium-based alloys was to use an ion-exchange membrane as an interface material which would gradually release minute concentrations of HCl vapour that then reacted with the liquid alloy surface and prevented residue formation. The exchange membrane, Nafion, was shown to act as an effective HCl transport medium and as a source of HCl vapor. It was possible to integrate commercially available extruded Nafion tubes or to use a suspension of Nafion to coat any suitable silicone-based microfluidics matrix material. Both methods were effective in producing a reversible flow system that remained residue-free for long periods and which could be regenerated without using concentrated acidic solutions.

In further work[21], the highest advancing angle was 175° and receding angle was 163° for a surface which was patterned with micropillars (175μm pitch distance between pillars) that were textured for additional roughness. The driving force which was required in order to actuate a 3μl galinstan droplet in the 3-dimensional super-lyophobic channel was 3.11mN.

Treatment with HCl vapour can also restore the non-wetting characteristics of galinstan[22]. It was shown that the oxidized (Ga_2O_3, Ga_2O) surface of the alloy was replaced by $InCl_3$ and $GaCl_3$. The surface tension and static contact-angle of HCl-treated galinstan on Teflon-coated glass were 523.8mN/m and 152.5°. Droplet-bouncing tests also demonstrated the non-wetting nature of HCl-treated galinstan. In a much more elaborate recent study[23], the spreading, recoiling and rebounding behaviours of galinstan droplets at metal-foam surfaces were determined. A high-speed camera was used to capture the droplet shapes. In the early stages of droplet impact, the spreading of droplets having a high surface tension was proportional to the square root of normalized time. This was consistent with that of conventional liquids and was related to the non-dimensional pore size of the foam surface during subsequent spreading. The maximum spreading factor of the galinstan droplets on small non-dimensional pore-size foam surfaces was greater than that on a smooth nickel surface, and decreased with increasing non-dimensional pore size. During rebound, the shape oscillation exhibited three modes due to differences in the pore structure. In one mode there was regular oscillation in the horizontal and vertical directions. In other modes, there was oscillation in the horizontal and vertical directions with rotation, or rotation oscillation. The usual theoretical formula for predicting the oscillation frequency of droplets was extended to cases involving irregular oscillations in droplet shape.

Contact-angle measurements have been made[24] of galinstan on various silicon surfaces over a range of temperatures. Due to the oxidation problem, this was done in an inert nitrogen environment containing less that 0.5ppm of oxygen. The contact-angle depended upon the substrate type and the surface structure, but the temperature had no discernible effect upon the contact angle. The latter ranged from 141° on smooth silicon, to more than 160° on silicon micropillars.

Galinstan (gallium–indium–tin), and the rival EGaIn (eutectic gallium–indium) alloy, exhibit similar flow behaviors, which are attributed to a thin oxide shell. It was found[25] that the interfacial tensions of galinstan/water and galinstan/1M-HCl were similar, at about 530mN/m. In 0.001 to 0.5M HCl, the interfacial tension decreased to 160mN/m. A similar behavior was found for the interfacial rheology. A low interfacial tension, together with a mechanically strong interface at intermediate acid concentrations, suggested that there was a change in interface composition. The SnO_2 particles which were present during dispersion produced more stable galinstan dispersions than did Ga_2O_3 or In_2O_3. The SnO_2 also improved the dispersion of EGaIn alloy, in spite of the absence of tin. Droplets of both of these alloys can incidentally be actuated by using the electro-wetting on dielectric effect. Application of 80 to 100V across the actuation electrode and ground electrode activates the droplets.

The interfacial properties of galinstan have also been measured[26] in a nitrogen-filled glove-box at 28C, with oxygen and moisture levels below 0.5ppm, in order to avoid the oxidation problem. It was noted that galinstan droplets would behave like a liquid only if they were never exposed to oxygen levels above 1ppm. The advancing and receding contact-angles were determined for sessile droplets on various substrates. In the case of glass, they were 146.8° and 121.5°, respectively. The surface tension, as measured by using the pendant-drop method, was 534.6mN/m. The corresponding angles[27] for Teflon were 161.2° and 144.4°.

The gallium alloy forms a thin surface oxide layer in the presence of a higher oxygen concentration. This oxide shell, unlike the bare liquid metal, sticks easily to the surface of almost any material, thus making it difficult to pattern. The oxide layer can be destroyed by treatment with acids or bases, and the bare liquid alloy below can wet the surfaces of certain materials. Contact-angle measurements were made of a Ga-In-Sn droplet before and after treatment with NaOH (figure 3). The contact angle for polydimethylsiloxane was the same after NaOH treatment, but it decreased sharply from 130 to 27° for gold films deposited on glass or polydimethylsiloxane substrates.

Another approach[28] is to use sand-blasting, chemical etching or coatings to make common surfaces non-wettable by galinstan. This concept is based upon its high surface tension and yield strength; factors which prevent the penetration of liquid metal into surface features. The resultant surfaces resemble traditional super-hydrophobic ones, and exhibit good non-wettability by galinstan; as indicated by high static and dynamic contact angles, little hysteresis and impact resistance. Reported fabrication method based on sandblasting, etching and spray coating is easily applicable to various surfaces ranging from metals, ceramics, to plastics and is scalable to large surfaces.

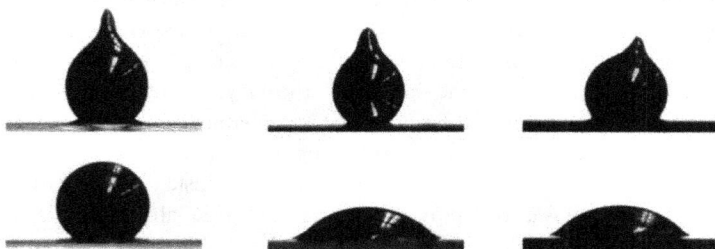

Figure 3. Change in contact angles of Ga-In-Sn droplets on polydimethylsiloxane (left), gold deposited on glass (centre) and gold deposited on polydimethylsiloxane (right). The upper images are those before treatment with sodium hydroxide, and the lower images are those following treatment with a 10wt% solution of sodium hydroxide. Reproduced under creative commons licence from: A skin-attachable, stretchable integrated system based on liquid GaInSn for wireless human motion monitoring with multi-site sensing capabilities, Y.R.Jeong, J.Kim, Z.Xie, Y.Xue, S.M.Won, G.Lee, S.W.Jin, S.Y.Hong, X.Feng, Y.Huang, J.A.Rogers, J.S.Ha, NPG Asia Materials, (2017) 9, e443

It is somewhat ironic that the oxidation of galinstan, which causes so many problems, produces an oxide which is useful in its own right. The wide-bandgap surface oxide of galinstan has been used as a template for the sonochemical synthesis of ultra-thin functionalized Ga_2O_3 nanosheets possessing tunable properties. In order to modulate the surface properties of Ga_2O_3 nanosheets, ionic aqueous solutions of $AgNO_3$ and $SeCl_4$ have been used for synthesis, given that decoration of Ga_2O_3 nanosheets with silver and selenium nanostructures aids the visible-light responsivity of the nanosheets.

Functionalization of the gallium oxide nanosheets was also associated with a unique rectifying behaviour. The formation of Schottky or ohmic junctions between a gold electrode and Ga_2O_3 nanosheets, and the creation of type-II heterojunctions at semiconductor hetero-interfaces, were controlling features of opto-electronic devices.

One of the earliest applications of the alloy was as a so-called femtosyringe which permitted femtolitre cum attolitre amounts of liquid to be injected into minute biological entities[29]. The heat-induced expansion of galinstan within a glass syringe was used to drive liquid through a tip having a diameter of about 0.1μm. This tip size caused less damage than did rival methods, while the heat-induced expansion permitted better control of the injection-rate. As a demonstration, standard staining agents could be injected into cyanobacteria and into sub-units of more complex organisms leading, for instance, to the injection of genetic material into a cyanobacterium so as to develop improved antibiotics. This feat was later dwarfed, oxymoronically speaking, by a pipette which was shown[30] by transmission electron microscopy to be capable of delivering molten metallic alloy with zeptolitre resolution. This was used to produce almost free-standing $Au_{72}Ge_{28}$ drops which were suspended from an atomic-scale meniscus at the pipette tip. This permitted phase transformations to be observed with near-atomic resolution. These observations suggested, incidentally, that liquid-state surface faceting could precede surface-induced crystallization and in turn by-pass nucleation in the interior. This use of the alloys is exploited up to the present time in the context of antennae.

A spherical helix structure, based upon 3-dimensional printing and microfluidic techniques was applied[31] to drug delivery, with liquid metal being introduced in order to increase the sensitivity of the reconfigurable helix topology. The influence, upon the resonant frequency of the structure, of the dimensions of the hemisphere which was traced out by the helical turns was investigated. The microfluidic channels of 1mm radius were arranged in the form of a double semispherical helix by using a 3-dimensional printer. In order to manage the liquid metal injection effectively, a syringe pump was incorporated in order to control the fluid flow flexibly. The device exhibited an excellent radiation gain and impedance matching, with a sensitivity of 0.96ml and 0.33ml/GHz; corresponding to frequency-shifts of from 2.28 to 2.01GHz and from 2.01 to 1.65GHz.

A galinstan electrode was found[32] to be suitable for reducing several functional groups in organic substances. The electrochemical behaviour of N-containing voltammetric-active drugs such as 1,4-benzodiazepines (e.g. nitrazepam) nitrofurantoin and phenazopyridine could be studied by using a liquid electrode in the form of a hanging galinstan drop. Concentrations of copper, cadmium, lead, bismuth, antimony and thallium at a level of 10^{-5} to 10^{-8}mol/l could be measured, and differential pulse and adsorptive stripping voltammograms could be recorded in various electrolytes, such as 0.1M KNO_3 and acetate or phosphate buffer solutions. The hanging galinstan drop electrode was able to accumulate metal ions at the electrode surface, thus making possible the simultaneous determination of two metals[33]. When using an acetic buffer solution, the potential window ranged from -900 to 150mV, while the limit of detection ranged from 6ppm for Sb^{3+} to 2ppb for Pb^{2+}. A renewable galinstan silver-based film electrode has subsequently been developed[34] for the voltammetric determination of lead, bismuth and thallium ions in acetate buffer solutions. A drop of galinstan is here placed in a sensor body which is filled with a saturated solution of sodium sulfite. The galinstan film is mechanically refreshed by pulling silver wire through the galinstan chamber, and then pushing it out of the sensor body. This provides very good surface reproducibility and permits Pb^{2+}, Bi^{3+} and Tl^+ determination at the micromolar concentration level.

Liquid electrodes prove to be useful in other contexts. Electrode polarization, for example, is an obstacle to the impedance measurement of ionic liquids. An atomically smooth electrode could reduce unwanted impedance contributions arising from electrode polarization. Liquid-metal electrodes have been formed by combining galinstan with acrylic plates in a parallel-plate capacitor arrangement[35]. The impedance of salt and protein solutions was then measured between 40Hz and 110MHz. Because of oxide-layer formation however, the results deviated greatly from ideal behaviour. The use of liquid electrodes for micro-electro-discharge machining has recently been proposed[36] in order to avoid problems related to electrode wear. Galinstan was supplied via a metallic capillary nozzle, such that the surface of the liquid projected above the flat bottom of the nozzle. A discharge pulse generator was then connected between the liquid, via the nozzle, and the work-piece. As the nozzle-tip, covered with liquid-electrode material, approached the work-piece, discharge pulses occurred when the gap satisfied breakdown-potential conditions and thermally removed the target material. During machining, liquid was continually supplied to the metallic micro-capillary nozzle in order to avoid any effect of liquid consumption upon target removal.

Returning to the medical field, the mercury thermometer has largely been phased-out in hospitals, in favour of thermocouple or infra-red body-temperature measurement. The relative accuracy of digital, mercury and galinstan thermometers has nevertheless been compared[37]. No statistically significant difference in mean temperature was found between mercury and galinstan thermometers in pair-wise comparison tests, but appreciable differences were found between mercury and digital, and between galinstan and digital readings. The sensitivity and specificity of digital thermometers were 67.5 and 98.0%, respectively. For body-temperatures greater than 39C, the false-negative rate was 65.4% for digital thermometers and 30.8% for galinstan thermometers. It was concluded that, although both digital and galinstan thermometers offered good specificity and positive predictive value as compared with those of a mercury thermometer, the galinstan thermometer possessed a higher sensitivity and gave a lower rate of false-negatives. A galinstan thermometer is therefore more accurate for the measurement of body temperature than are digital or mercury thermometers.

The eutectic galinstan naturally tends to attack many of the metals which are used as electrical contacts and connectors in electronics. For instance, when galinstan is in direct contact with aluminium thin films, the latter is easily dissolved and this leads to the formation of aluminium oxides on the surfaces of galinstan droplets[38]. In the presence of a multilayer graphene diffusion barrier however, the aluminium-galinstan reaction is effectively prevented when galinstan is deposited by drop casting. When deposited by spray-coating, the high impact effect of galinstan droplets damages the multilayer graphene and aluminium-galinstan reaction occurs at damaged locations. A graphene barrier is nevertheless likely to block the formation of aluminium oxides at the aluminium/galinstan interface.

Galinstan also finds applications in the field of catalysis. An easy method, for the preparation of catalytically active materials from galinstan liquid, is sonication in an alkaline solution or treatment in a reducing medium[39]. This creates solid In/Sn-rich microspheres which exhibit catalytic activity.

The so-called galvanic replacement reaction is a useful method for the preparation of bimetallic nanomaterials[40]. The galvanic replacement of liquid galinstan with platinum produces mainly Pt_5Ga. During this process, a plume of nanomaterial may be ejected upwards from the centre of the liquid metal droplet and into solution, while hydrogen gas is liberated because the platinum-rich nanomaterial is a very effective catalyst for hydrogen evolution. The Pt_5Ga is very effective in the electrochemical oxidation of methanol and ethanol. It was deduced that the increased activity was due to the anti-poisoning properties of the surface with regard to CO upon incorporating gallium atoms into a platinum catalyst. The galvanic replacement reaction is also useful for creating nanostructured materials. The galvanic replacement of liquid galinstan with silver and gold can be achieved by using a macro-size droplet to create a liquid-metal marble which consists of a liquid core and a solid shell in which the form of the outer shell is determined by the concentration of metallic ions which existed in solution during the galvanic replacement process. This permits the possible recovery of precious metal ions from solution: in metallic form and immobilized on the liquid metal, thus facilitating recovery[41].

Galinstan can be used as a simple and low-toxicity way to synthesize high surface-area alumina[42] by first slowly dissolving aluminium, something which it does all too readily in the electronics context. The gallium-indium-tin-aluminium alloy is then selectively oxidized, at room-temperature and pressure, under a damp stream of flowing CO_2 or N_2. This yields amorphous alumina having a high surface area and very low water-content, with the surface area ranging from 79 to $140m^2/g$, depending upon the initial percentage weight of aluminium used. The as-synthesized alumina aerogel appears to be blue, due to Rayleigh scattering from its fibrous dendritic nanostructure. Upon annealing at 850C, the amorphous product transforms into crystalline Al_2O_3. The relationship between the initial aluminium concentration in the alloy, and the surface area of alumina, peaks at about 30%Al.

Gallium-based liquid metals have even been essayed[43] as lubricants, by adding them to grease in various proportions and homogenizing the mixture via mechanical agitation and ball-milling. The tribological properties were then judged by using the four-ball test. This revealed that pure gallium-based liquid metal exhibited a poor anti-wear performance under low loads, while the lubricating performance of greases under high loads was also unsatisfactory. When the liquid metal was added to the grease, the extreme-pressure lubricating capacity of the latter increased while the anti-wear performance under low loads did not decrease. When the mass ratio of molten metal to commercial lithium grease reached unity, the weld-load was over 10kN; considered to be the maximum value for greases.

The transient hot-wire method was modified[44] by using a flexible U-shaped quartz capillary filled with galinstan. This was then used to measure the thermal conductivity of molten KNO_3–$NaNO_3$–$NaNO_2$ eutectic. This probe was able to measure the thermal conductivity of the molten salt, showing that it ranged from 0.48W/mK at 500K, to 0.47W/mK when close to 700K. The quartz retained its electrical non-conductivity, and there was no detectable current leakage over the above temperature range.

Although the coolant properties of liquid metals, and the prospect of replacing toxic mercury, are primary concerns in liquid metal research, the purpose of the present work is to present their usefulness in those fields where their remarkable ability to flex and extend, without the risk of fracture or fatigue which afflicts solid metals, is invaluable. Room-temperature liquid metals such as galinstan have the useful ability to maintain electrical and thermal conductivity during deformation. One example of such an application is that of strain sensors. A soft silicone elastomer and a liquid metal alloy such as galinstan can be used to make an electrofluidic strain sensor which can handle deformations greater than 200% and cycle repeat-times in the hundreds. Another application, to be covered in great detail later, is that of mechanically-controlled frequency-agile antennae; the agility being obtained by using a liquid metal. Galinstan-based ink-jet printing can be used to create highly-stretchable electronic devices which possess moreover a self-healing capability. An LED-integrated circuit having such a self-healing capability can survive being twisted through more than 180° or being stretched by up to 60% more than 2000 times.

A pressure-conductive rubber sensor, comprising a liquid-metal plus polydimethylsiloxane composite, could match the curvature of the human body and measure finger movements, blood pressure and breathing[45]. The composite was made by mixing galinstan and polydimethylsiloxane using magnetic stirring. This composite was conductive only when the mechanical pressure or strain exceeded a threshold level; the composite being otherwise non-conductive. This threshold value could be controlled by choosing the ratios of liquid metal and polydimethylsiloxane. The material could monitor elastic deformation, produced by pressing, stretching or bending, without suffering degradation or structural failure. The use of an array of such sensors could reveal the distribution of applied pressure over a plane.

EGaIn

It was pointed out some time ago[46] that EGaIn is well-suited, to the applications to be described here, because of its room-temperature rheological properties. That is, it behaves like an elastic material, up to a critical surface stress, beyond which it yields and begins to flow easily. This is important because it often has to be injected into microfluidic channels in order to form stable microstructures, and its properties permit it to fill microchannels rapidly when sufficient pressure is applied to the channel inlets. At the same time, it can maintain structural stability within the channels when ambient pressure is restored. Experiments which were carried out in microfluidic channels, and in a parallel-plate rheometer, suggested that its behavior is governed by the properties of its mainly gallium oxide surface layer. Both experimental techniques indicated about the same magnitude (0.5N/m) for the critical surface stress which was required in order to cause EGaIn to flow. Further analysis showed that the pressure which had to be exceeded, so that EGaIn would flow through a microchannel, was inversely proportional to the smallest dimension of the channel. Gallium-oxide coating is also a method for rendering essentially any flat surface non-wetting with respect to naturally-oxidized gallium-based liquid metal alloys. Materials such as polydimethylsiloxane, bare silicon, SU-8 on silicon, SiO_2 layers on silicon and glass have been tested[47] with regard to contact angle, bouncing and rolling behaviour with pneumatically driven galinstan droplets on gallium oxide coated flat surfaces being moved at 1.45 to 3.07cm/s. This simple method did not necessitate micro/nano fabrication or nanoscale surface topology in order to avoid stiction of naturally-oxidized gallium-based liquid metals and was applicable to common microfluidic substrates.

Stable suspensions of EGaIn nanoparticles can be formed via probe-sonication of the metal in aqueous solution[48]. Positively-charged molecular surfactants, such as cetrimonium bromide or lysozyme, stabilize the suspension by interacting with the negative charges of the surface oxide that forms on the metal. The liquid metal breaks up into nanospheres during sonication, but form rods of gallium oxide monohydroxide under moderate heating in solution, during or after sonication (figure 4). An interesting point is that, whereas heating normally drives a phase transformation from solid to liquid, it here instead drives the transformation of particles from liquid to solid, via oxidation. Indium nanoparticles also form during shape transformation, due to the selective removal of gallium. This de-alloying phenomenon provides a mechanism for creating indium nanoparticles at temperatures which are well below its melting point. It was shown to be possible to transform and de-alloy other gallium alloys, including ternary ones.

Nanoparticles made of non-noble metals currently attract attention due to their promising applications in ultra-violet plasmonics, realised so far in the form of pure solid or liquid gallium particles immobilized on solid substrates. Colloidal liquid-metal nanoparticle solutions are however more important, and strong ultra-violet plasmonic resonances of EGaIn liquid-metal nanoparticles suspended in ethanol have been demonstrated[49]. Most relevant here is their value to the use of plasmonics in stretchable electronics.

The nano-particles exhibited ultra-violet plasmonic resonances which agreed with effective medium models for a 100nm particle, the main resonances being located at 213 and 275nm. The results were relevant to the concept of reconfigurable liquid-metal plasmonics, in which the plasmon resonances of liquid-metal nano-particles can be tuned by deforming the shapes of the nano-particles by using means such as ultrasound. Although the plasmonic properties of liquid-metal nano-particles made from pure gallium are similar, the lower (circa 15.5C) melting-point of EGaIn as compared with that (circa 30C) of gallium itself makes it more suitable for room-temperature applications. Mie theory-based simulations of gallium and cesium liquid metals, with their core–shell nanospheres coated with silver, were again used[50] to investigate the localized surface plasmon resonance wavelengths, and the sensitivity of the peak position to particle-size and shell-thickness. It was concluded that the absorption and scattering efficiencies of the liquid metal nanoparticles, coated with silver, could be tailored by controlling the size of the core and shell layers. Resonance peaks were found in the ultra-violet, visible and near-infrared regions in the case of bare nanoparticles, while they were found in the ultra-violet and visible regions in the case of core–shell nanoparticles. The absorption efficiency increased with increasing core size or shell thickness.

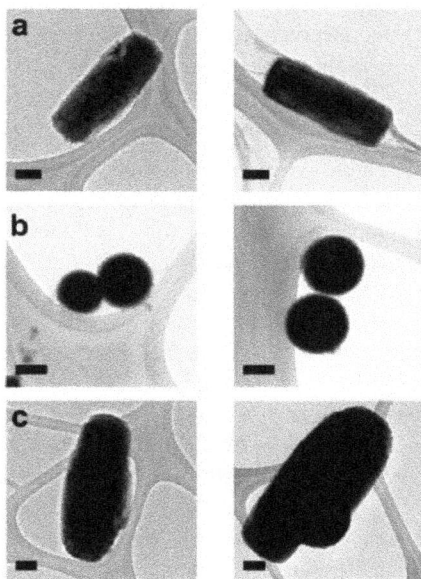

Figure 4. Transmission electron microscopic images of liquid-metal nanoparticles: a, rods obtained by sonication of gallium (left) and EGaIn (right) in aqueous solution containing Lys protein, b, nanospheres synthesized by repeating previous treatment in an ice bath, c, rods obtained by heating the previous samples. The black bars are 100nm long. Reproduced under creative commons licence from: Shape-transformable liquid metal nanoparticles in aqueous solution, Y.Lin, Y.Liu, J.Genzer, M.D.Dickey, Chemical Science, 2017, 8, 3832.

Adaptive wireless devices have used the movement of EGaIn within microchannels as an electrical switching mechanism, thus avoiding the problems posed by the use of older solutions such as diodes and micro-electromechanical systems. Reconfiguration using EGaIn is however greatly limited by the occurrence of permanent short-circuits which are caused by retention, of the liquid in the microchannels, due to wetting and rapid surface-oxide formation. Recent investigations[51] have targeted those conditions which impair repeatable electrical switching when using EGaIn in microchannels. Initial contact-angle tests of EGaIn on epoxy surfaces have demonstrated its wettability of flat surfaces. Scanning electron microscopic cross-sections of microchannels have revealed the

Materials Research Forum LLC

https://doi.org/10.21741/9781644900697

adhesion of EGaIn residue to the channel walls, and micro-computed tomographic scans have furnished volumetric measurements of the amount of EGaIn which remains within channels following flow cycling. Future progress is expected to require the use of non-wetting coatings.

Gallium-indium alloys other than galinstan have been applied to the cooling of integrated circuits and electronic chips. When water was used as a coolant, microchannels were integrated into device substrates in order to manage heat fluxes of 1 to 4W/cm². Two methods were later proposed[52] for improving the thermal-control performance of substrates which used gallium-indium alloy as the working fluid in channels with a width of 400μm, and with via holes located in the area of the microchannels. A comparison with the cooling performance of de-ionized water for devices with a heat flow density of 1 or 2W/cm² showed that, when the heat flux was increased to 5W/cm², the cooling performance of the de-ionized water was not ideal, while the metal was able to lower the maximum temperature of the substrate to 354K when it was flowing at a rate of 70ml/min and the heat flux was increased to 10W/cm². By adding via-holes to the substrate, the maximum temperature could be reduced to 357K, with the alloy flowing at 70ml/min while the heat flux density was 30W/cm². The heat-transfer and hydrodynamic behaviours of liquid metal in a vascular-like microchannel-based cooling system were also compared[53] with that of water as a coolant. The microchannel networks comprised 2 vascular systems which were vertically and mutually connected. Each system had a single inlet or outlet channel, and bifurcated uniformly over many levels. Such vascular networks, combined with liquid metal alloy, could constitute a high-efficiency thermal management system for electronic devices.

An attempt was made[54] to clarify the heat-transfer behavior of liquid metal flowing in mini-channel exchangers having various geometrical configurations. The liquid metal alloy, Ga-20In-12%Sn, was used as a coolant at various mass flow rates in 3 forms of simulated heat-exchanger. All of the simulated heat sources were cooled by the 3 heat-dissipation devices and the exchanger having the smallest channel-width exhibited the highest mean heat-transfer coefficient under all conditions. This was attributed to its much larger heat-transfer area.

Although the present work is devoted to the use of gallium-indium based alloys in electrical engineering fields, it is worthwhile mentioning some of their close relatives and their other applications.

A magnetocaloric ferrofluid has been based[55] upon a gadolinium-saturated liquid-metal matrix by using a gallium-based alloy as the solvent and suspension medium. The room-temperature liquid material exhibited spontaneous magnetization and a large magnetocaloric effect; the magnetic properties being attributed to the formation of gadolinium nanoparticles, suspended within the liquid gallium alloy which acted as a reaction solvent during nanoparticle synthesis. Nanoparticle weight-fractions which were greater than 2% could be suspended within the liquid matrix. This ferrofluid was a promising material for use in magnetocaloric cooling due to its high thermal conductivity and its fluidity. The formation of nanometre-sized metallic particles within the supersaturated liquid metal solution was expected to be useful for chemical synthesis and for the preparation of metallic nanoparticles based upon highly reactive rare-earth metals.

Liquid metal has been modified[56] by using magnetic copper–iron nanoparticles, such that the modification permitted the nanoparticles to be suspended within the liquid metal. The liquid then exhibited a similar appearance, actuating behavior and deformability, in alkaline solutions, to those of the pure liquid-metal alloy. The modification permitted however the rapid and precise control of the liquid by means of a magnetic field. Precise control and climb motion of liquid metal was demonstrated, for the first time, by simultaneously applying electric and magnetic fields.

Table 2. Conversion efficiency of gallium-tin alloys as extreme ultra-violet light sources

Sn (at%)	Melting Point (C)	Conversion Efficiency (%)*
0	29	0.5
10	22	0.8
20	30	1.0
30	30	1.3
40	30	1.3
100	232	1.3

* 13.5nm, 2% band-width

A simple method for preparing surface-enhanced Raman-scattering substrates, by using sonochemical treatment and mechanical stirring without surfactants, has been proposed[57] for the detection of molecules at low concentrations. Liquid metals were used as surface-enhanced Raman-scattering substrate materials, and were caused to form an oxide film; thus leading to a good dispersion of the nanoparticles. So-called nanograss, rod-like structures and nanogaps, formed on the micro/nanoparticle surface and provided the surface-enhanced Raman-scattering sites. The shape and size of the particles could be modified by adjusting the ultrasonication time and the stirring speed. The behaviour of a surface-enhanced Raman-scattering substrate which was coated with gold film was evaluated by using rhodamine-6G as a probe. The limit of rhodamine-6G Raman detection, using an optimized nanograss-structured substrate, could be as low as $10^{-7}M$; with a standard deviation of 8 to 15.5%.

Liquid gallium-tin alloy can act as a laser target source for 13.5nm light generation. These alloys melt at about 30C and are therefore at the upper limit of the sort of alloy considered here. A Nd:YAG laser (1064nm, 1ns, 7.1 x $10^{10}W/cm^2$) was used[58] to ablate the alloy, thereby generating a similar extreme ultra-violet emission intensity which was similar to that of tin, in spite of the relatively small quantity of tin which was present. Data indicated that a 30at% tin content was required to give an estimated melting point of about 30C and conversion efficiency of 1.3%, like that of bulk tin (table 2).

Surface-sensitive X-ray scattering techniques with atomic-scale resolution were used long ago incidentally in order to investigate the structure of the surface of various classes of liquid binary alloys and the surface segregation predicted by the Gibbs adsorption rule[59]. In gallium-indium alloy, the first layer consisted of a supercooled indium monolayer while the bulk composition was reached only after about 2 atomic diameters.

Because of the relatively high cost of indium, other alloys have been proposed as being a cheaper alternative to galinstan. One of these alloys is based upon gallium, tin and zinc. This ternary system is bounded, on a 3-dimensional phase diagram, by 3 simple eutectic systems. The Ga–Sn–Zn eutectic, with a composition comprising 6.64at% of tin and 3.21at% of zinc, has a melting point of 288K. Touch-printing can be used to obtain nanometric films from the alloy surface of the alloy. One method permits the transfer of oxidised material, on the liquid metal surface, to other substrates. The method has been used to obtain 2-dimensional Ga_2O_3 films. The oxide layer which formed on Ga–Sn–Zn alloy was transferred onto other substrate, while the contact angle between the liquid alloy and the substrate was also determined[60]. The magnitude of the contact angles did not affect the deposition of the oxide layers, and it was possible to obtain nanometric oxide layers which were just a few μm in size. The oxide layer contained some 90at% of gallium, plus some tin and zinc, and appeared to be nanocrystalline in nature. The contact angle between Ga–Sn–Zn eutectic and silicon, glass and quartz substrates were 134°, 132° and 118°, respectively. The values which were obtained in the case of silicon were similar to that (139°) for galinstan on silicon at 303K, but galinstan was non-wetting on some substrates, such as glass, when not oxidised. As soon as the oxide skin on galinstan was broken however, the alloy stuck to the surface and gave the impression that it wetted the substrate. Thin films which were obtained from eutectic Ga–Sn–Zn alloy contained not only gallium oxide, but also smaller quantities of tin and zinc oxides, and it was unclear whether the tin and zinc had replaced the gallium in the oxide or were present in the form of small patches of tin oxide and zinc oxide among the gallium oxide.

Noting that galinstan is opaque to wavelengths of 325 to 850nm, while polydimethylsiloxane is highly transparent over the same range, a real-time dynamically reconfigurable photomask was developed[61] which used liquid metal as an ultraviolet-opaque material in ultraviolet-transparent polydimethylsiloxane. Conventional ultra-violet lithography uses a fused silica plate photo-mask covered with chromium. Bright-field and dark-field microfluidic photomasks could be used here to transfer various patterns onto a positive photoresist.

Gallium-based liquid metals are not only non-toxic, but also exhibit good biocompatibility. When they are used in cancer treatment however, their high surface tension leads to unstable nanoparticles. When ZrO_2 is coated onto liquid-metal nanoparticles they form a stable core-shell nanostructure and liquid-metal nanoparticles, coated with ZrO_2, retain the favorable flexibility which is beneficial for cellular uptake. The coated liquid metal nanoparticles rapidly warm up and emit the desired amount of heat under near-infrared laser irradiation. The coated nanoparticles are also more effectively accepted by cells, and are beneficial in the field of tumour photothermal therapy. EGaIn particles can also deliver anti-cancer drugs to tumours when they have their shapes changed by external means. Suitable particles can be prepared by sonication of solutions which contain EGaIn and the amphiphilic lipid, 1,2-distearoyl-sn-glycero-3-phosphocholin. These composite particles can then transport an anti-cancer drug such as doxorubicin, which is released when the particles are deformed by light or heat[62].

Eutectic gallium-indium alloys such as EGaIn can be made into nanoparticles which have a passivating gallium oxide shell, with a wide range of possible thicknesses, by using various chemical treatments[63]. Physical deformation of the particles can then release the conductive molten contents from the oxide shell. The mechanical properties of the core-shell liquid nanoparticles were related to the shell thickness by using nano-indentation to measure the particle stiffness and elastic modulus. Nano-indentation studies also revealed the onset of particle rupture, and of a resultant conductivity. Particles having a gallium oxide shell thickness of between 1.28 and 4.46nm were obtained, and the mechanical properties were measured by nano-indentation. The results revealed particle-stiffness values that ranged, with increasing shell-thickness, from 0.05 to 2.9N/m (figure 5). It was suggested that the presence of ligands during oxide growth modified the modulus in some way which was not directly related only to the oxide shell thickness. Permitting the oxide to 'mature' (figure 6) seemed to lead to a more homogeneous rigid structure containing fewer defects. Shells which were treated with a combination of thiols and phosphonic acids during initial oxide growth were thinner and were thus more likely to be heterogeneous. Conductive-tip indentation tests revealed the critical force required for particle rupture and hence the release of the liquid EGaIn. There was a several-orders-of-magnitude difference in the forces required for particle rupture, as judged by the appearance of conductivity between tip and substrate.

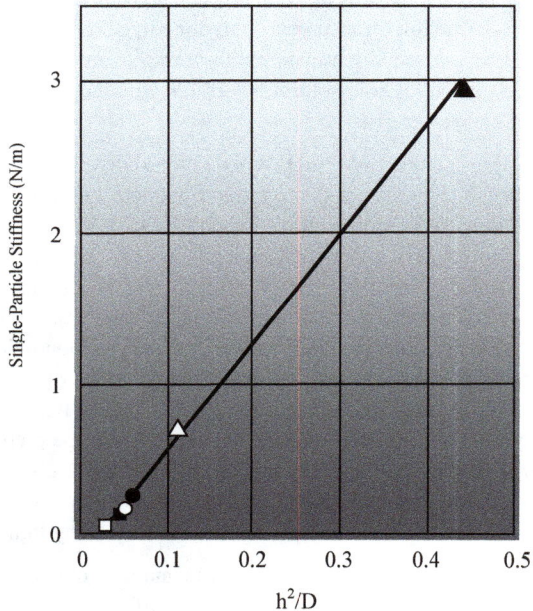

Figure 5. Single-particle stiffness of EGaIn-containing nanoparticles as a function of the square of the shell thickness divided by the particle diameter. Open square: 1-dodecanethiol treatment, closed square: 2,3,4,5,6-pentafluorothiophenol/4-fluorobenzyl phosphonic acid treatment, open circle: dodecylphosphonic acid treatment, closed circle: untreated, open triangle: 11-phosphonoundecanoic acid treatment, closed triangle: H₂O₂/phenol treatment

Electromechanical actuators can exploit the Lorentz force in order to convert electrical energy into rotational or linear kinetic energy, but the electrical current flows through wires which are rigid; thus limiting the types of motion which are possible. The latest advances in the preparation of liquid metals provide the possibility of wires that are far more flexible. A new type of motor which uses liquid-metal conductors has been proposed[64], in which radial modes are activated. Micro- and nano-motors can move, and fulfil other functions, at small scales; with that usefulness depending upon novel materials such as liquid metals, with their associated electrical conductivity, biocompatibility and flexibility. The construction of motors made from galinstan and having dimensions as small as a few hundred nanometers has been reported[65]. The motors were half-coated with a thin platinum layer, and moved in H_2O_2 due to self-electrophoresis. The motors can move within a silver nanowire network; moving along the nanowires and accumulating at the contact junctions where they become fluidic and lead to junction microwelding at room temperature by reacting with acid vapor. They thus constitute nanorobots which are capable of repairing circuits by small-scale welding.

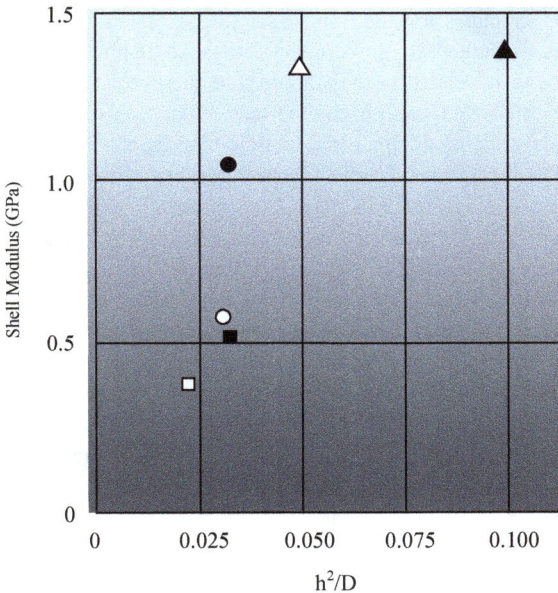

Figure 6. Calculated shell modulus as a function of the square of the shell thickness divided by the particle diameter. Open square: 1-dodecanethiol treatment, closed square: 2,3,4,5,6-pentafluorothiophenol/4-fluorobenzyl phosphonic acid treatment, open circle: dodecylphosphonic acid treatment, closed circle: untreated, open triangle: 11-phosphonoundecanoic acid treatment, closed triangle: H_2O_2/phenol treatment

A class of electrostatic actuators, involving a compliant electrode made from liquid metal contained within a thin elastomer, has been used as on-chip microvalves for gas-flow control. The microvalve consisted[66] of a fixed electrode which spanned the bottom and side-walls of a trapezoidal gas channel, plus a corresponding flexible electrode which was suspended across the channel.

One of the main roles that has traditionally been played by liquid metals is that of a coolant; a familiar example being the use of molten sodium-potassium alloys in nuclear reactors. It is no surprise then that alloys which are liquid even at room temperature have found applications in this domain. The use of liquid metal as a thermal-interface material has been considered as being a possible means of solving the problem of combining contact heat-transfer with a high heat flux. The effect of liquid-metal wettability upon contact heat transfer was studied[67] by depositing silver, nickel, molybdenum or tungsten onto copper plates via direct-current magnetron-sputtering. Galinstan was then used to sandwich the copper plates together, and laser flash analysis was used to measure the thermal diffusivity, conductivity and contact resistance of the sandwich structures. The thermal properties were related to the surface free energy of the buffer layer materials. With increasing surface free energy, the thermal contact resistance decreased while the thermal diffusivity and the thermal conductivity increased. Samples with a tungsten buffer-layer could have a thermal contact resistance which was as low as $0.402 mm^2 K/W$, with a thermal conductivity of 378.261K. As compared with samples without buffer layers, the thermal contact resistance could be decreased by about 64.05%. In earlier work[68], the structure and wettability of molybdenum-coated plate had been characterized by means of X-ray diffractometry and drop-shape analysis. In the former results, molybdenum and copper peaks had been observed, with no other compounds being detected. The contact angles of water on the copper plate and the molybdenum-coated copper plate were 65.9 and 17.8°, respectively, showing that the wettability of the molybdenum-coated copper plate was superior. The thermal contact resistance of copper|copper samples was $937.095 mm^2 K/W$, with a thermal conductivity of 3.596W/mK. The thermal contact resistance of copper|liquid|copper samples was $1.119 mm^2 K/W$, with a thermal conductivity of 350-356W/mK. Samples with a molybdenum layer could have a thermal contact resistance which was as low as $0.616 mm^2 K/W$, with a thermal conductivity of 369-484W/mK. Following wettability modification by coating, the thermal contact resistance decreased by about 45%, indicating that the molybdenum buffer layer played an essential role in increasing heat transfer in the liquid|copper interface-region. An attempt was made[69] to increase the interface heat-transfer of galinstan still further by adding copper particles. As usual, the liquid metal alloy was introduced between copper plates and laser flash analysis was used to measure the overall thermal conductivity. The addition of the copper particles clearly improved the thermal performance of the oxidized liquid metal alloy. The thermal conductivity and thermal contact resistance of oxidized liquid metals, with particle mass fractions of 5 and 10%, were 200.33 and 233.08W/mK and 7.955 and 5.621mm^{2}K/W, respectively. As compared with oxidized liquid metal without copper particles, the thermal conductivity had been increased by about 68 and 96%, and the thermal contact resistance decreased by about 57 and 70%, respectively. The fluidity of the liquid metal was reduced by the addition of copper particles. A highly-conductive thermal paste was also described which again consisted of liquid metal alloy and copper particles. Galinstan was here mixed with copper particles which had a average diameter of 9µm. During particle dispersion, de-gassing was used to remove air bubbles and further improve the thermal conductivity of the paste. A laser flash technique was used to measure the thermal conductivity. Three types of pastes were compared: liquid metal alloy, oxidized liquid metal alloy and oxidized liquid metal alloy plus 5wt% of copper particles. The thermal conductivities could attain 44.48, 13.55 and 24.34W/mK, respectively, with corresponding thermal contact resistances of 4.044, 5.638 and 4.075mm^{2}K/W. When the fraction of copper particles was increased from 5 to 10wt%, the thermal conductivity increased from 24.34 to 29.07W/mK and the thermal contact resistance decreased from 4.075 to 3.37mm^{2}K/W.

The organic molecular compounds known as plastic crystals have often been used to model solidification phenomena[70,71], and so it is not surprising that room-temperature liquid metal alloys have been used to model those aspects of solidification modelling which do not expressly benefit from transparency of the analogue, or where conductivity plays a critical role. An experimental set-up has thus been used[72] to model the continuous casting of steel, operated continuously in a loop arrangement with Ga-In-Sn alloy as the steel analogue. The set-up comprised a round mould, with an inner diameter of 80mm, which represented the mould used in industrial bloom-casting, and an electromagnetic-stirrer which generated a rotating magnetic field having various magnetic-flux densities. The velocity profiles within the mould were measured via ultrasound Doppler velocimetry. The results revealed a surprising behaviour of the submerged jet in that, in medium magnetic field strengths, the jet ceased to circle and stayed in one position in the mould. In a real system, this could well have risked mould failure due to the continuous impingement of superheated liquid on a single point.

Flow control which is based upon time-dependent magnetic fields is used in various processes involving liquid metals, and so a suitable measurement technique for flow-mapping in model experiments is the above ultrasound Doppler velocimetry. Unlike conventional systems which use transducers having a fixed sound field, another approach[73] is to use a phased array which has the ability to focus and steer an acoustic field. Another non-contact method for flow-rate control is time-of-flight Lorentz force velocimetry, and this determines the flow rate via measurement of the Lorentz force which acts on magnetic pick-ups which are placed close to the flow. A vortex generator is used to generate an eddy in the flow, with two magnet systems of known separation placed downstream of the vortex generator. Each magnet system has a force sensor which can detect the passing of the eddy through the field as an appreciable perturbation. The flow-rate is then deduced from the time delay between the perturbations registered at each sensor. When galinstan was electromagnetically pumped round a closed rectangular loop, in a refinement of the method[74], 3-dimensional strain-gauge force sensors were used to measure the Lorentz force resulting from flow disturbances in various directions. A closely-related method, again based upon the induction of eddy currents using a time-harmonic magnetic field and subsequent measurement of the resultant secondary magnetic field using gradiometric induction coils, can be used to determine the position and topology of the surface of a liquid metal. The method has been applied[75] to the static and dynamic surfaces of galinstan held in a narrow vessel. A precision of better than 1mm, and a time-resolution of at least 20Hz can be achieved.

Both methods, local Lorentz-force velocimetry and ultrasound Doppler velocimetry, have more recently been combined[76] in order to perform velocity measurements in a vertical turbulent convection flow-cell filled with galinstan. This demonstrated the applicability of local Lorentz force velocimetry to thermal convection flow, and revealed a linear dependence of the measured micronewton force upon the magnitude of the local flow velocity. The experiment explored the scaling-laws of the global turbulent transport of heat and momentum in low Prandtl-number convection flow. Due to the reciprocity of these Lorenz-force effects, transient eddy currents in galinstan which is situated above an excitation coil can be determined[77] by measuring the corresponding voltage-drops using an electrical-potential probe. The resultant spatio-temporal eddy-current field data then provide useful checks on the accuracy of analytical expressions. As an example, the disturbance of the eddy-current distribution due to the presence of a non-conducting torus immersed in the liquid metal, can be measured and compared with numerical results.

Room-temperature liquid metal alloys are in general very convenient for studying such magnetohydrodynamic phenomena. When a liquid metal is placed in a horizontal cylinder with a vertical magnetic field, and there is an additional magnetic field, an induced current is produced by convective movement of the liquid metal. The induced current and magnetic field interact to produce a Lorentz force which opposes the movement and inhibits natural convection and thus the heat-transfer intensity. Liquid gallium, in particular, exhibits a deceleration of the flow velocity. A magnetic field in the horizontal radial direction, which is perpendicular to the natural convection flow caused by the temperature gradient, has a significant effect upon natural convection, whereas the effect upon the axial direction is comparatively weak in a moderate magnetic field. The inhibition becomes more obvious with increasing intensity of the magnetic field. Because such effects can occur at any scale, they can be used as analogies for even celestial phenomena. Magnetorotational instability, for example, plays a primordial part in the formation of stars and black holes by enabling outward angular momentum transport in accretion discs. The use of combined axial and azimuthal magnetic fields, as above, therefore permits the investigation of stellar effects in easily handled liquid-metal flows at moderate Reynolds and Hartmann numbers. The use of galinstan in this way[78] has provided experimental evidence for the occurrence of magnetorotational instability in Taylor-Couette flow.

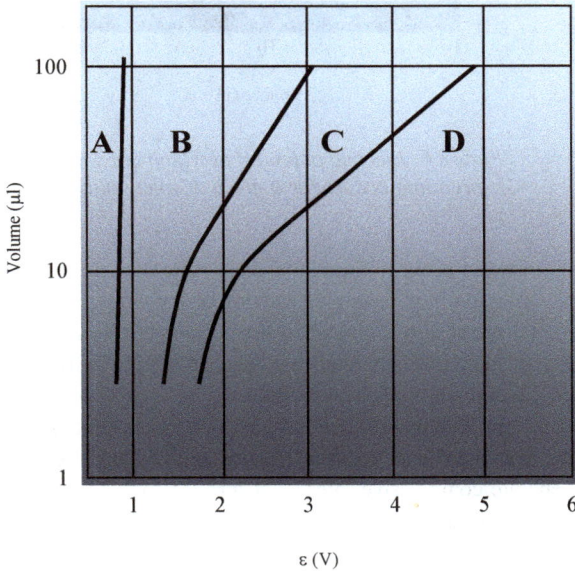

Figure 7. Phase diagram for droplets of volume, V, subjected to an electric potential, E. A: smooth sessile droplets, B: fractal morphology, C: undulated spreading followed by contraction, D: spreading followed by contraction.

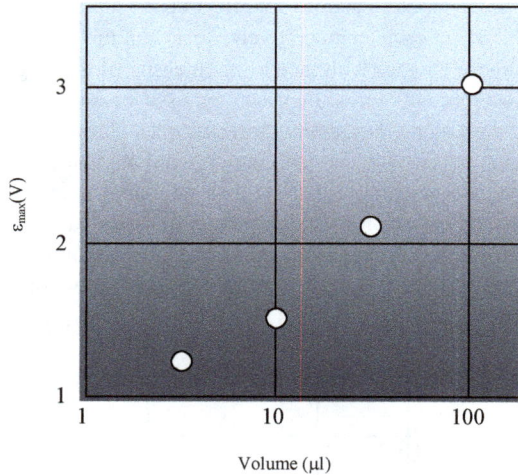

*Figure 8. Electric potential at which maximum
spreading occurs for a given droplet volume*

Fingering instabilities can occur in liquid metals in spite of their high interfacial tensions. Electrochemical oxidation, for example, can lower the effective interfacial tension of gallium-based liquid metal alloys almost to zero. This permits the occurrence of almost any interface shape, including a fractal one. Shapes having a fractal dimension of 1.3 have been observed, and study of their morphology as a function of droplet volume and applied electric potential has revealed which 3 main factors govern such interface behaviour. These are: interfacial tension, gravity and oxidative stress. Electrochemical oxidation can generate compressive interfacial forces which oppose the tensile forces at the liquid interface. The layer of surface oxide constitutes a physical limitation, and an electrochemical barrier, which blocks instability at larger positive potentials. An extensive study[79] (figures 7 to 10) has been made of the balance between interfacial tension and oxidative compressive stresses at an interface due to the importance of shape-control in many applications.

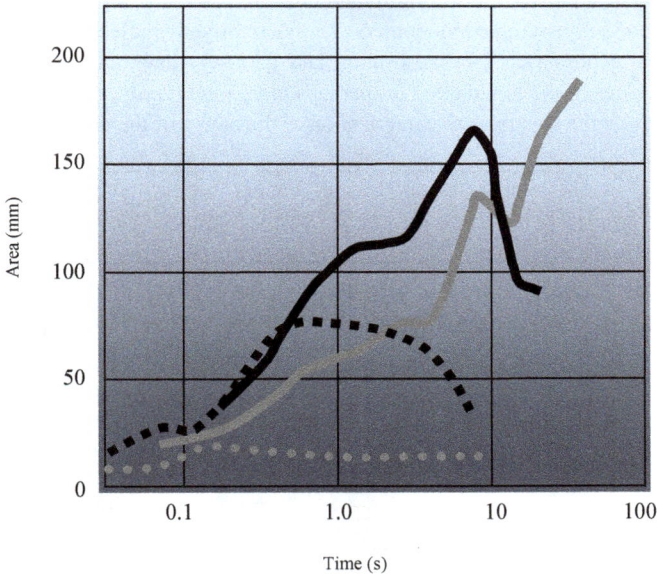

Figure 9. Area of spreading droplet as a function of time. Region A: grey dotted line, region B: grey solid line, region C: black solid line, region D: black dotted line.

Self-Assembled Monolayers

This is a prime example of the manner in which liquid metals can be used as the very softest of metallic contacts for the gentle probing of the electrical properties of delicate systems. The thin oxide layer does not interfere with the study of sensitive electrical phenomena, such as charge transport through self-assembled monolayers, because the latter tend to be considerably more resistive than is the oxide. Multiple contacts, of controlled geometry, can moreover be established by injecting the metal into microchannels which are placed on the substrate of interest.

Charge transport across large-area molecular tunnelling junctions is widely studied because of its potential importance in the development of quantum electronic devices, and large-area junctions which are based upon gallium-indium eutectic – as used in the form of a conical-tip top-electrode - have become a reliable test-bed for exploring structure-property relationships. Gallium-indium eutectic has found a particularly useful role in the study of self-assembled monomers. These are organic molecular layers which spontaneously form on surfaces by adsorption and generally have an ordered domain structure. The surface of the liquid alloy or, perhaps more importantly, the oxide layer on the alloy surface permits electrical characterization of the nature of the monolayer.

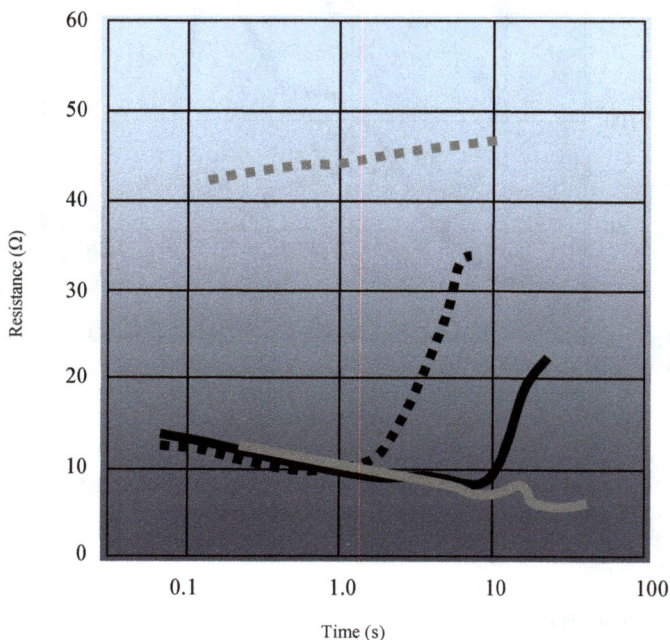

Figure 10. Resistance of spreading droplet as a function of time. Region A: grey dotted line, region B: grey solid line, region C: black solid line, region D: black dotted line.

In a typical early study, the current-density versus applied bias-voltage behaviour of self-assembled monolayers of 3 ethynylthiophenol-functionalized anthracene derivatives was studied[80]. They had essentially the same thickness, but exhibited linear, cross or broken conjugation. The liquid eutectic offered a supporting native skin, about 1nm thick, of Ga_2O_3 to act as a non-damaging conformal top-contact. The skin also imparted non-Newtonian rheological properties. The current-densities for the broken and cross conjugated molecules were not very different: with a current density of $0.1A/cm^2$ at a bias voltage of 0.4V. On the other hand, in the case of linear conjugation the current density was one or two orders of magnitude higher than for the others. These values were in good qualitative agreement with theoretical calculations, for single molecules chemisorbed between gold contacts, that also predicted higher values for the linear case. The overall deduction was that a linearly conjugated molecule was much more conductive than was either a cross-conjugated or broken conjugate. The latter could also be difficult to differentiate. The results were attributed to quantum interference effects, and suggested that molecule-molecule interactions did not play a great role regarding the transport properties.

Tunnelling junctions of the form, Ag^{TS}-$S(CH_2)_{n-1}CH_3 \| Ga_2O_3/EGaIn$ where TS signifies 'template-stripped', permitted the study[81] of charge transport across the self-assembled monolayers. Under ambient conditions, the surface of the liquid Ga-24.5wt%In electrode (melting-point: 15.7C) oxidizes and also adsorbs other surface contaminants. The interface between the EGaIn and the self-assembled molecules thus includes the oxide film, plus any adsorbed organic material. The interface naturally affects the properties and behaviour of the junction. In the present case, the oxide was about 0.7nm thick, was composed mainly of Ga_2O_3 and appeared to be self-limiting in growth. The structure and composition of the junctions were conserved from junction to junction, and the transport of charge through the junctions was dominated by the alkane thiolate self-assembled monolayer rather than by the oxide or any contaminants. The interface between the oxide and the eutectic alloy was rough at the μm-scale. Replacing a -CH_2CH_2- group with a -CONH- group changed[82] the dipole moment and polarizability of a portion of the molecule and might conceivably have changed the rate of charge transport through the self-assembled monolayer. In fact, it had no great effect upon the rate of charge transport across junctions of the form, Ag^{TS}-$S(CH_2)_mX(CH_2)_nH \| Ga_2O_3/EGaIn$, where X was -$CH_2CH_2$- or -CONH-. On the other hand, incorporation of the amide group increased the fraction of non-shorting junctions; as compared with n-alkane thiolates of the same length. This suggested that combining a thiol group at one end of a molecule, with a test-group at the other while using an amide-based coupling, could identify molecules which would be useful in the study of molecular electronics. Analysis of the rates of tunnelling across self-assembled monolayers of n-alkane thiolates, SC_n, where n is the number of carbon atoms in the chain, when incorporated into junctions of the above form, indicated[83] a value of $103.6A/cm^2$ (V = +0.5V) for the injection tunnel current density. That is, the current which flows through an ideal junction with n = 0. This estimate did not involve a length-extrapolation, because it was possible to measure the current densities across self-assembled monolayers for lengths ranging from n = 1 to n = 18. This estimate also assumed that the geometrical contact area equalled the effective electrical contact area. More detailed experimental study revealed however that the roughness of the Ga_2O_3 layer and that of the Ag^{TS}-self-assembled monolayer led to effective electrical contact areas that were some 10^{-4} times the corresponding geometrical contact areas. Re-calculation led to an injection tunnel current (+0.5V) of = $107.6 A/cm^2$. This was comparable to values which had been reported for junctions using top-electrodes of evaporated gold and graphene, and to the values which were estimated for tunnelling through single molecules. In these EGaIn-based junctions, the tunnelling decay factor of 0.75/Å or 0.92/nC fell within the ranges (0.73 to 0.89/Å or 0.9 to 1.1/nC) reported for various types of junction.

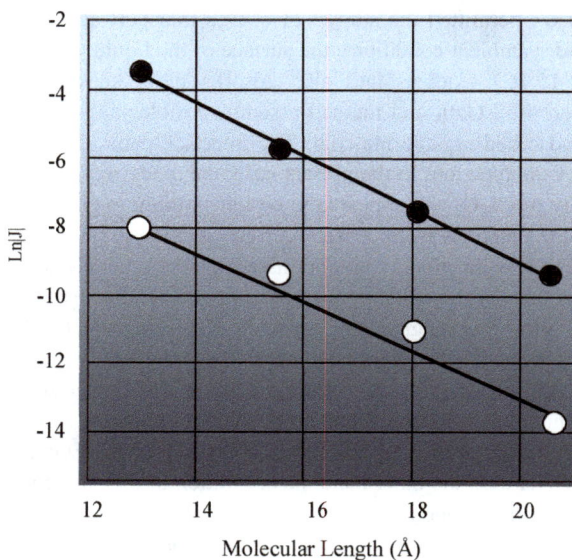

Figure 11. Injection current as a function of molecular length for Ag^{TS}/SAM||EGaIn junctions involving $CH_3(CH_2)_9SH$, $CH_3(CH_2)_{11}SH$, $CH_3(CH_2)_{13}SH$, and $CH_3(CH_2)_{15}SH$. Upper line: ambient conditions. Lower line: N_2 atmosphere with 1 to 3%O_2 and relative humidity below 15%

A comparison was made[84] of the current versus voltage characteristics of self-assembled monolayer-based tunnelling junctions, having top EGaIn electrodes, which had been prepared using 2 different methods. In one case, the electrode had been stabilized in elastomer microchannels. In the other case, the electrode had been suspended from the tip of a syringe. These two geometries of the EGaIn electrode, at least when in contact with air, furnished identical data. The junctions incorporated self-assembled monolayers of $SC_{n-1}CH_3$, with n = 12, 14, 16 or 18, which were supported on ultra-flat template-stripped silver electrodes. Both methods produced high (70 to 85%) fractions of junctions that were sufficiently stable to perform current-voltage measurements in statistically large (400 to 1000) numbers.

The samples with the top electrode stabilized in microchannels also made it possible to perform current-voltage measurements as a function of temperature (110 to 293K). The J(V) characteristics were independent of temperature, and were linear under low (-0.10 to 0.10V) bias. The current density decreased exponentially with increasing thickness of the self-assembled monolayer. The results suggested that tunnelling was the main mechanism of charge transport across the junctions. Both methods yielded tunnelling decay coefficients of about 1.0/nC (0.80/Å), and the pre-exponential factor (including contact resistance) was some $3.0 \times 10^2 A/cm^2$. It was concluded overall that the current-voltage characteristics of Ag^{TS}-monolayer‖Ga_2O_3/EGaIn junctions are dominated by the structure of the organic component of the monolayer and not by electrode type, surface-film resistance or metal work-function.

Most of the early studies of these systems had focused on saturated molecules having backbones which consisted mainly of alkanes in which the frontier orbitals were either highly-localized or energetically inaccessible. Self-assembled monolayers of wire-like oligophenylene ethynylenes, which are fully conjugated, exhibit length-dependent tunnelling behavior only in a low-oxygen environment. This anomalous behavior was attributed[85] to the sensitivity of the injection current to the environment and it was concluded that the self-limiting layer of Ga_2O_3 strongly affected the transport properties and that the effect was related to the wetting behavior of the electrode. It was further suggested that adhesive forces play a significant role in tunnelling charge-transport in large-area molecular junctions. Further research showed that the environments under which self-assembled monolayers and junctions of large-area Au^{TS}/SAM‖Ga_2O_3/EGaIn junctions with mono- and di-thiol oligophenylenevinilene, and Ag^{TS}/SAM‖Ga_2O_3/GaIn junctions with alkanethiolate self-assembled monolayers, formed has a marked effect upon tunnelling charge-transport. That is, the resistance of self-assembled monolayers with oligophenylenevinilenes decreased in low-O_2 low-humidity environments, while that of self-assembled monolayers with alkanethiolates increased. By comparing self-assembled monolayers of oligophenylenevinilenes which had either a bare phenyl group or a thiophenol group, to the EGaIn interface and self-assembled monolayers of

Liquid Metal Alloys in Electronics   Materials Research Forum LLC
Materials Research Foundations **70** (2020)   https://doi.org/10.21741/9781644900697

alkanethiolates under ambient conditions and a N_2 atmosphere with 1 to 3%O_2 and a relative humidity of less than 15%, the observed effects could be attributed to the nature of the SAM/Ga_2O_3 interface. The injection currents, but not the decay constants, were affected by the environment (figure 11). Variable temperature measurements showed that the transport mechanism through oligophenylenevinilenes, at low-O_2 and low relative humidity, was tunnelling. The wetting behaviour at the SAM||Ga_2O_3/EGaIn interface was a critical factor ... and could become a limiting factor in the case of π-conjugated molecules with small decay constants relative to those of n-alkanes.

A similar comparison was made[86] of the charge transport across junctions by tunnelling in the structures, Ag^{TS}-X(CH_2)$_{2n}CH_3$||Ga_2O_3/EGaIn, with n = 1 to 8 and with X= -SCH_2- or -O_2C-. The replacement of Ag^{TS}-SCH_2-R by Ag^{TS}-O_2C-R, where R is an alkyl chain, had no appreciable effect upon the pre-exponential factor (3 x 10^3A/cm^2) or tunnelling decay factor (0.75 to 0.79/Å). This again indicated that structural and electronic changes at the interface did not affect the rate of charge transport. A comparison of junctions involving oligophenylene carboxylates and n-alkanoates showed that the tunnelling decay factor for aliphatic (0.79/Å) and aromatic (0.60/Å) self-assembled monolayers differed markedly. It was proposed that silver self-assembled monolayer interfaces, with either a thiolate or carboxylate anchoring group, may be directly comparable.

A further study was made[87] of the effect of uncharged polar functional groups upon the rate of charge transport, by tunnelling, across self-assembled monolayer large-area junctions of the form, Ag^{TS}-S(CH_2)$_n$M(CH_2)$_m$T||Ga_2O_3/EGaIn, where M and T were middle and terminal functional groups. The 12 uncharged polar groups, CN, CO_2CH_3, CF_3, OCH_3, $N(CH_3)_2$, $CON(CH_3)_2$, SCH_3, SO_2CH_3, Br, $P(O)(OEt)_2$, $NHCOCH_3$, $OSi(OCH_3)_3$, with permanent dipole moments ranging from 0.5 to 4.5, were incorporated into the self-assembled monolayer. A comparison of the electrical characteristics of these junctions, with those of junctions formed from n-alkane thiolates again led to the conclusion that the rates of charge tunnelling were insensitive to the replacement of terminal alkyl groups by terminal polar groups in this set. The current densities measured here also suggested that the tunnelling decay parameter and injection current for self-assembled monolayers terminating in non-polar n-alkyl groups, and for polar groups selected from among common popular ones, are statistically indistinguishable. Introducing Lewis acidic/basic functional groups, T = -OH, -SH, -CO_2H, -$CONH_2$ or -PO_3H, into the structure, Ag^{TS}/S(CH_2)$_n$T||Ga_2O_3/EGaIn where, it will be recalled, S(CH_2)$_n$T is a self-assembled monolayer of n-alkane thiolate with a terminal functional group, T, makes it possible to determine the response, of rates of charge transport by tunnelling, to changes in the strength of the interaction between T and Ga_2O_3. This yielded values of tunnelling current density which were indistinguishable from the values observed for n-alkane thiolates of equivalent length[88]. This insensitivity of the tunnelling-rate to changes in the terminal functional group implied that replacing weak van der Waals contact interactions with stronger hydrogen or ionic bonds, at the $T\|Ga_2O_3$ interface, did not change the shape (height or width) of the tunnelling barrier to an extent which was sufficient to affect the rates of charge transport. A comparison of the injection currents for $-CO_2H$ and $-CH_2CH_3-$, groups of similar extended lengths, suggested that both groups made indistinguishable contributions to the height of the tunnelling barrier.

Odd-even effects were rarely observed in molecular junctions involving self-assembled monolayers of n-alkanethiolates, and it was unclear whether they were an interface effect or were caused by intrinsic properties of the self-assembled monolayers; or were a combination of both. Odd-even effects were therefore studied[89] in self-assembled monolayer-based tunnel junctions of the form, $AgA-TS-SC_n\|GaO_x/EGaIn$, with n = 2 to 18. Use of a combination of alternating- and direct-current techniques permitted the separation of interface effects from other contributions. An odd-even effect which was observed, in the value of J, by using direct-current methods was caused by the intrinsic properties of the self-assembled monolayers. Clear odd-even effects were observed in the values of the self-assembled monolayer resistance and self-assembled monolayer capacitance, but the odd-even effect in contact resistance was too weak to be responsible for an odd-even effect which was observed in the current densities. The odd-even effects in the present junctions were therefore attributed to the properties of the self-assembled monolayers and self-assembled monolayer-electrode interactions, which together determine the shape of the tunnelling barrier. In a more recent study[90] of the odd-even effect, a method was developed for chemically polishing the EGaIn electrode tip in order to permit the formation of smooth conformal contacts by re-establishment, of their liquid character at the point of contact, via tension-driven reconstruction of a thin oxide layer. In order to evaluate the polished tip, the charge-transport behavior across n-alkane thiolate self-assembled monolayers was measured, and a good correlation was found between the odd-even oscillation behavior and wetting-study data. Because the molecules are homologues of one another and differ only in the orientations of the terminal CH_2CH_3 component, the odd-even effect is governed by orientation-induced differences in the absence of self-assembled monolayer (gauche) defects. A comparison of the present data with previously published results revealed a clear difference between odd-numbered self-assembled monolayers with respect to silver and gold.

The current-rectification ratio, $J(-2.0V)/J(+2.0V)$, for supramolecular tunnelling junctions with a top electrode of EGaIn and its conductive 0.7nm supportive oxide layer, increases by up to 4 orders of magnitude under an applied bias of +1.0 to +2.5V[91]. Such junctions did not change their electrical characteristics when biased at ±1.0V. The increase in the current-rectification ratio is caused by the presence of water and ions in supramolecular assemblies. These react with the Ga_2O_3/EGaIn interface and increase the thickness of the Ga_2O_3 layer. This increase in the oxide thickness, from 0.7 to about 2.0nm, changes the nature of the monolayer/top-electrode contact from ohmic to non-ohmic. These results unambiguously set the experimental conditions, a safe bias window of ±1.0V, for the investigation of molecular effects in electronic junctions with Ga_2O_3/EGaIn top electrodes, where electrochemical reactions are not significant. The interpretation of data arising from studies involving biases greater than 1.0V may thus be complicated by electrochemical side-reactions. These can be revealed by changes in the electrical characteristics as a function of voltage-cycling.

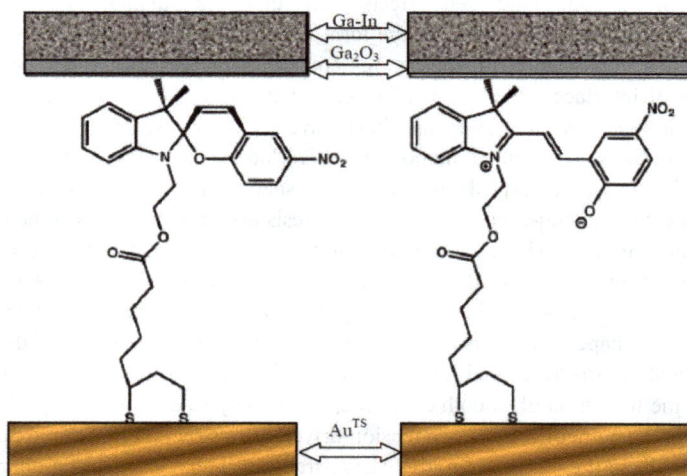

*Figure 12. Spiropyran self-assembled monomer in EGaIn/Ga_2O_3∥SAM/Au^{TS}
Left: closed form with length of 15.4Å, right: open form with length of 13.3Å*

The photo-induced switching of conductance in tunnelling junctions containing self-assembled monolayers of spiropyran has been studied[92] by using EGaIn top contacts. Although the spiropyran component was separated from the electrode by a long alkyl-ester chain (figure 12), there was an increase in the current density by a factor of 35 at 1V when the material was irradiated with ultra-violet light in order to induce a ring-opening reaction. The degree of switching of hexanethiol-mixed monolayers was higher than that of pure spiropyran monolayers. A first switching recovered 100% of the initial current-density level and, in the mixed-monolayers, subsequent damping was not the result of monolayer degradation. The observed increase in conductivity was supported by zero-bias calculations which predicted a change in a localization of the density of states near to the Fermi level, and by simulated transmission spectra which exhibited positive resonances that broadened and shifted toward the Fermi level when in the open form. The linking of stretchable materials to mechanochromic molecules also provides the ability to display direct cues[93]. For example, the spiropyran mechanophore can be covalently incorporated into polydimethylsiloxane so that a visible color-change signals that a strain-threshold has been reached. The coloured elastomers can also be moulded and patterned so that a direct warning such as $\boxed{\text{STOP}}$ appears[94]. The strain which triggers the appearance of colour can moreover be tailored by layering silicone polymers having differing moduli. Colour-warnings can also be used to indicate when a certain frequency has been reached by a stretchable liquid-metal antenna.

The ambient atmosphere which is used in these studies also plays a role, as shown[95] by using a gas-tight chamber to control the atmosphere in which the electrodes were formed, and in which measurements were made of the current densities across self-assembled monolayer-based junctions. Atmospheres of air (plain or containing acetic acid or water vapour), oxygen, nitrogen, argon and ammonia were combined with rough conical-tip electrodes or smooth hanging-drop electrodes: so called because handling of the oxide film during creation of the former electrodes led to appreciable μm-scale roughness at the electrode surface while the extrusion of drops created a much smoother surface. Both geometries were used to compare junctions across a self-assembled monolayer of n-dodecanethiol. In air, this gave $\log|J|_{mean} = -2.4$ for a conical tip and $\log|J|_{mean} = -0.6$ for a drop electrode. This increase in current density was attributed to a change in the effective electrical contact area of the junction. Junctions which comprised a graphite electrode and a hanging-drop electrode were compared in experiments where the electrodes did, or did not, have a surface oxide film. The presence of the oxide then clearly did not affect measurements of $\log|J(V)|$, and thus did not contribute to the electrical resistance of the electrode. On the other hand, the presence of an oxide film improved the stability of junctions and increased the yield of working electrodes from some 70% to about 100%. Increasing the relative humidity (20 to 60%) in which $J(V)$ was measured did not affect the results for (CH_3)- or (CO_2H)-terminated self-assembled monolayers.

In general, the behavior of these metals is directly related to the thickness of the surface oxide layer. The latter can be determined non-destructively, using ellipsometry, if the dielectric coefficient is known. The dielectric responses (which closely obeyed the Drude model) of liquid gallium and of gallium-indium eutectic alloy have been determined[96] at room temperature, from 1.24 to 3.1eV, by means of spectroscopic ellipsometry. Overlayer-induced artefacts were eliminated by applying a reducing potential to the metal surface while immersed in an electrolyte. The data for EGaIn suggested that, in the absence of the oxide, the surface was indium-enriched.

The tunnelling rates in many test situations were observed to decrease exponentially, with a decay coefficient that was close to 1.0/nC, across n-alkanethiolate monolayer-based tunnelling junctions. It was unclear however why there was such a large scatter, of up to 12 orders of magnitude, in the injection current-densities; this being the hypothetical current density which flows across the junction when n = 0. Every type of junction contains defects or impurities in the electrode materials, and the presence of such defects is a key factor causing an increase in observed values of injection current. The number of defects in $Ag^{TS}/SC_n||GaO_x/EGaIn$ junctions was adjusted[97] by varying the geometrical contact-area of the junction. The value of the current (about $100A/cm^2$) was independent of the junction size when the geometrical contact-area was small ($<960\mu m^2$), but increased by 3 orders of magnitude, from 10^2 to $10^5 A/cm^2$, when the geometrical contact-area was increased from 969 to $18000\mu m^2$. With increasing current, the yields in non-shorting junctions decreased from 78 to 44% while the decay coefficient increased from 1.0 to 1.2/nC. The junction quality could be qualitatively assessed by observing the curvatures of dJ/dV plots. That is, the presence of defects changed the sign of the curvature from positive, which was associated with tunnelling, to negative; which was associated with Joule heating. It was concluded that the electrical behaviour of large junctions was governed by thin-area defects while that of small junctions was governed by the molecular structure. More recent research[98] has again addressed the relationship between the rate of charge transport (by tunnelling) across self-assembled monolayers in a metal/self-assembled-monolayer$||Ga_2O_3/EGaIn$ junction, and the geometrical contact area between a conical $Ga_2O_3/EGaIn$ top electrode and the bottom electrode. Measurements of the current density, J(V), across self-assembled monolayers of decanethiolate on silver showed that J(V) increases with geometrical contact area when the contact area is small ($<1000\mu m^2$), but reaches a plateau at between 1000 and $4000\mu m^2$. The method which is used to prepare $Ga_2O_3/EGaIn$ electrodes produces a tip for which the apex is thicker and rougher than its thin smoother sides. When the geometrical contact area is small, the $Ga_2O_3/EGaIn$ electrode thus contacts the bottom electrode mainly at the rough apex and forms inconsistent areas of electrical contact. When the geometrical contact area is large, contact occurs via the smoother regions around the apex and this is a much more reproducible situation. Measurements which were made of the contact pressure, existing between conical EGaIn electrodes and atomic force microscope cantilevers, showed that the effective contact pressure, as governed by the mechanical behavior of the oxide skin, decreased roughly inversely as the diameter of the geometrical contact area. Such self-regulation of the pressure thereby prevented damage to the self-assembled monolayer and maintained the ratio of the electrical contact area to the so-called geometrical footprint at an approximately constant value. In the field of molecular electronics in general, it is critically important to minimize the number of causes that can produce defective electrodes. These include contaminants such as photo-resist and organic residues which are involved in the preparation process, or roughening of the electrode during etching. These defects impair the creation of well-organized molecular structures. Junctions which were based upon micropores would be desirable because the latter are scalable. Micropores are not produced on ultra-smooth template-stripped electrodes however, and may produce stray capacitances and leakage currents across the insulating matrix. Micropores can be produced in AlO_x on template-stripped gold-based arrangements in such a way that the gold surface is not in direct contact with photo-resist during preparation. Such junctions therefore do not exhibit capacitance or leakage problems and remain stable for months. The quality of such junctions can in fact be best judged by comparing them with junctions involving cone-shaped tips of EGaIn, or with EGaIn held in a hole in polydimethylsiloxane.

A detailed study has been made[99] of the van der Waals interface which forms between indium-tin oxide and a liquid metal micro-electrode, particularly with regard to the contamination of the oxide surface by airborne contaminants. Large-area junctions of the form, ˙indium-tin-oxide‖Ga_2O_3/EGaIn were studied, where Ga_2O_3 was the usual 0.7nm-thick self-passivating oxide film on EGaIn and ‖ represented the van der Waals interface which formed between indium-tin oxide and the EGaIn electrode. Current-density data were collected for junctions formed at indium-tin oxide surfaces that had been prepared using various surface-cleaning methods. The electrical conductance of the indium-tin-oxide‖Ga_2O_3 van der Waals interface was markedly enhanced (3 orders of magnitude, as judged by comparing the median values of the log-current density) by means of ultra-violet/ozone surface-cleaning. The scatter in the data was also significantly reduced.

Very recent experiments[100] have been aimed at testing the effect, of terminal groups which contain carbon-halogen bonds, upon the current density (via hole-tunnelling) across self-assembled monolayer-based junctions of the form, M^{TS}/$S(CH_2)_9NHCOCH_nX_{3-n}$‖Ga_2O_3/EGaIn, where M is silver or gold and X is CH_3, fluorine, chlorine, bromine or iodine. These tunnelling-rates are found to be insensitive to the type of terminal group at the interface between the self-assembled monolayer and the Ga_2O_3, and such results thought to be relevant to the cause of the apparent ongoing inconsistency concerning the perceived influence of halogen atoms, at the self-assembled-monolayer‖electrode interface, upon the tunnelling current density.

The most recent research[101] has focussed on the relationship between molecular structure and the rectification of the tunnelling current in junctions which are based upon self-assembled monolayers, as in $Ag-S(CH_2)_{n-1}CH_3\|GaO_x/EGaIn$ junctions, where n is 10, 14 or 18, and in molecular diodes of the form, $Ag-S(CH_2)\|ferrocene\|GaO_x/EGaIn$, and how this depends upon the contact area between the self-assembled monolayer and the cone-shaped EGaIn tip. Large junctions, in which the contact area is greater than $1000\mu m^2$, are unreliable and defects such as pinholes govern the charge-transport characteristics. In the case of $S(CH_2)\|ferrocene$ self-assembled monolayers, the rectification ratio decreases from 130 to unity with increasing contact area, due to an increase in the leakage current: the current which flows across the junction at reverse bias when the diode ostensibly blocks current flow. The value of the decay coefficient decreases from 1.00 to 0.70/nC with increasing contact area; again signalling that large junctions suffer from defects. Small junctions, having contact areas of less than $300\mu m^2$, are not stable due to the high surface tension of the bulk EGaIn, which leads to unstable EGaIn tips. The contact area of such small junctions is dominated moreover, as noted previously, by a rough tip-apex which markedly reduces the effective contact area and reproducibility. The contact area of very large junctions, as also noted before, is controlled by the relatively smooth side-walls of the tip. It was found here that there was an optimum range, for the contact area, which lay between 300 and $500\mu m^2$ and where the electrical properties of the junction were governed by molecular effects. Within this range the value of current density increased with contact area until it reached a plateau for junctions having contact areas greater than $1000\mu m^2$, as previously reported. Molecular dipoles, chosen from among simple organic functional groups, have lately[102] been introduced into junctions having the structure, $Ag^{TS}/S(CH_2)_nR(CH_2)_mCH_3\|Ga_2O_3/EGaIn$, where R is an n-alkyl fragment such as $(-CH_2-)_2$ or $(-CH_2-)_3$, an amide group such as -CONH- or -NHCO-, an urea group, (-NHCONH-) or a thiourea group, (-NHCSNH-). These amide, urea or thiourea groups then introduced a localized electric dipole moment into the self-assembled monolayer and changed the polarizability of *that* section of the self-assembled monolayer but did not produce large electronically-delocalized groups or change any other feature of the tunnelling barrier. The local change in the electronic properties could be correlated with a small but significant rectification of the current. This demonstrated that the simplest form of current rectification, at ±1.0V, in EGaIn junctions is an interfacial effect which is due to a change in the work-function of the self-assembled monolayer-modified silver electrode resulting from the proximity of the dipole that is associated with the amide or other group. It was therefore not due to a change in the width or mean height of the tunnelling barrier.

A comparison had been made[103] of the electrical properties of self-assembled monolayers formed, on template-stripped gold, of 2 homologous series of 5 different oligophenylenes with alkane thiol tails. The terminal phenyl ring was replaced by a 4-pyridyl ring in one series, so that the 2 series differed only by the substitution of C-H for N. Tunnelling junctions were formed by, as usual, exploiting the liquid eutectic as a non-damaging conformal top contact which was insensitive to functional groups. Conductance measurements alone could not differentiate sufficiently between the 2 series of molecules. The length-dependences of the series of self-assembled monolayers produced tunnelling decay coefficients of 0.44 and 0.42/Å for pyridyl- and phenyl-terminated self-assembled monolayers, respectively. These values lay between the expected values for alkane thiolates and oligophenylenes. The values of the transition voltage were about 0.3V higher for the phenyl-terminated, than for the pyridyl-terminated, self-assembled monolayers. A comparison of the values of the current, with the highest occupied molecular orbital levels determined by density functional theory calculations, revealed an odd-even effect for the phenyl-terminated monolayers but not for the pyridyl-terminated monolayers. Graphs of the transition voltage versus the shift in work function were roughly linear. Discrepancies can arise due to differing tip-morphologies and fabrication methods, and it can be problematic to draw reliable conclusions on the basis of molecular features. One particular problem is a discrepancy, the so-called odd-even effect, between the behaviors of hydrocarbons containing odd and even numbers of carbon atoms. Graphs of the difference between the highest occupied molecular orbital and the Fermi energy were here clearly linear, with 2 distinct series that starkly differentiated the 2 series of monolayers. This was true even for self-assembled monolayers having nearly identical highest occupied molecular orbital levels, but only when the dipole-induced shift in vacuum level was considered. The effect of the electronic properties of the monolayers emphasized the importance of molecular dipoles in tunnelling junctions comprising self-assembled monolayers. Tunnelling junctions which incorporated EGaIn as a top contact were clearly sensitive enough to differentiate between self-assembled monolayers that differed by the substitution of a single atom. It has recently been demonstrated[104] that the behavior of self-assembled molecular monolayers can be modified by varying the composition and structure of the molecular layer using mixed monolayers. Such mixed monolayers were prepared on gold, having a surface roughness of about 1nm, via the co-adsorption of 11-(ferrocenyl)-undecanethiol as a rectifier and 1-undecanethiol as a diluent. Micron-sized molecular junctions were formed by using indium gallium eutectic as the top electrode, showing that the ratio of 11-(ferrocenyl)-undecanethiol to 1-undecanethiol molecules could affect the rectifying effect of the monolayer device. That is, the higher the fraction of ferrocene the better was the rectification. A mixed-monolayer consisting of 20% 1-undecanethiol and 80% 11-(ferrocenyl)-undecanethiol had a better rectification ratio than that of a pure 11-(ferrocenyl)-undecanethiol monolayer. This was due to a reduced leakage current. The presence of loosely-packed molecules on the surface of the pure 11-(ferrocenyl)-undecanethiol monolayer was due to the bulky head-group of the 11-(ferrocenyl)-undecanethiol compound and the roughness of the gold substrate. That part of the 11-(ferrocenyl)-undecanethiol which lay partly on the surface or which was buried in the layer created defects which were at the origin of leakage currents. When 1-undecanethiol molecules were inserted between the ferrocene molecules, the molecules in the monolayer became more ordered, as reflected by a decreased wave-number of the C-H stretching-mode of the methylene group. An increase in the thickness of 80% 11-(ferrocenyl)-undecanethiol monolayers, relative to that of pure 11-(ferrocenyl)-undecanethiol monolayers, implied that there existed a better orientation of the 11-(ferrocenyl)-undecanethiol molecule in the mixed monolayers. The ordered structure and superior orientation greatly improved the stability and reproducibility of the molecular devices, reduced the leakage current and increased the rectification ratio.

Energy-Harvesting

This is a very popular field of research at the present time, due to worries over environmental threats such as global warming and the need to find new non-polluting sources of energy. It often edges very close to pseudoscience however. It has its roots in the older concepts of regenerative braking and electromagnetic damping of motor-vehicle oscillations, where the recuperated energy would otherwise be truly lost. Some designers appear to ignore Newton's third law however and forget that their device actually imposes an added load, no matter how small, on the machine or person from which the energy is harvested. Using pedestrian movement to power street-lights is fine in principle, but not if the harvesting method makes those pedestrians feel as if they are trudging through swamp. Harvesting energy from soldiers' boots is fine, as the harvesting means might feel no different from sponge rubber, but clothing which harvested energy could lead to additional fatigue.

Given these caveats, liquid metal alloys are certainly promising materials for the design of energy-harvesting means, provided that the additional weight of metal to be carried is not a significant factor.

Due to their ability to flex and stretch, liquid metal interdigital electrodes which are encapsulated in an elastomer[105] constitute a novel means for skin-mounted tribo-electric energy-harvesting. The great advantage of liquid metals which are embedded in elastomers is that failures can be easily fixed and that the geometry of the interdigitated electrode can be rearranged simply by cutting and gluing. Such electrodes can be used for capacitive sensing on solid or liquid objects having various geometries which are accessed with greater difficulty by non-elastic electrodes. The elastomer-embedded interdigital liquid-metal electrodes can be mounted on human skin and can generate energy when fingers tap or 'caress'; a difference in solicitation which the electrode can detect. Tribo-electric nanogenerators which are based upon interdigitated elastomer-embedded liquid-metal electrodes can be used as elastic membranes which transform mechanical work into electrical energy when vibrated. Due to the previously mentioned ease of 'cutting and pasting', a two-dimensional tribo-electric generator can be cut into pieces and reassembled in three-dimensional form, thus offering a new route to building three-dimensional tribo-electric nanogenerators.

It has been proposed[106] that a liquid-metal based tribo-electric nanogenerator could have Galinstan as the electrode and silicone rubber as the tribo-electric encapsulant. The small Young's modulus of the liquid metal means that the electrode remains continually conducting under deformation and can be stretched to strains of up to 300%. An oxide layer on the galinstan prevents the liquid electrode from being further oxidized and from permeating the silicone rubber. Using a single-electrode form at 3Hz, a tribo-electric nanogenerator having an area of 6cm x 3cm could produce an open-circuit voltage of 354.5V, a transferred short-circuit charge of 123.2nC, a short-circuit current of 15.6μA and an average power density of 8.43mW/m^2. These parameters tended to remain stable during stretching, folding and twisting. Such material, when worn as a bracelet or clothing, could clearly harvest energy from normal human activities without fatiguing the wearer any more than does the familiar self-winding watch.

Ensuring adequate stretchability of these materials has been a problem. A previous nanogenerator[107] which was based upon fractal-like piezo-electric nanofibers and liquid-metal electrodes could sustain strains of up to only 200%. Large-scale fractal polyvinylidene fluoride micro/nanofibers were fabricated by combining helical electrohydrodynamic printing with buckling-driven self-assembly. The printing method exploited the so-called whipping and buckling instability of the electrospinning process to deposit snake-like fibers of various shapes in a controlled manner. The self-organizing buckling allowed the driving force of the pre-strained elastomer to arrange the fibers into an ultra-stretchable fractal-like architecture. The nanogenerator, with its embedded fractal fibers and liquid-metal microelectrodes, exhibited a stretchability of above 200% and an electrical output of more than 200nA.

In another attempt[108], polymers were used as tribo-electric layers, together with plastic metal films as a conductive layer, in order to produce an electrode having high flexibility. One polymer film was prepared so as to have a microstructural array which enhanced friction. The plastic metal was prepared by mixing liquid gallium–indium alloy with a glaze powder having a good coating ability, extensibility and conductivity. The resultant flexible electrodes exhibited a bending-angle of better than 180°, a radius-of-curvature of less than 1mm and a stable conductivity. The tribo-electric generator could produce an average voltage of 80V and a current of 37.2μA. A more recent design[109] employs a soft malleable porous composite of liquid metal and elastomer in which the random pores of the composite act as minute tribo-electric nanogenerators. The tribo-electric foam then produces a maximum peak-to-peak short-circuit current of about 466nA, a charge density of about 35μC/m^2 and an open-circuit voltage of some 78V within a sample size of 5cm x 5cm x 1cm. Surface texturing further increased the foam softness and increased charge generation: a 36% augmentation in tribo-electric charge to a charge density of about 48μC/m^2. During jogging, a porous shoe insert produced an instantaneous power of some 2.6mW at an instantaneous power density of 13μW/cm^2.

It was shown theoretically some time ago that a vibrating pair of parallel electrodes, bridged by a deformed liquid metal droplet, could harvest vibrational energy. The Young-Laplace equation was solved numerically for all of the possible droplet shapes taken up during a vibration cycle and the resultant time-dependent capacitance served as the input for an equivalent circuit model. When the results were applied to 2 existing energy-harvester designs, it was found that the optimum electrode separation had to be as small as possible or as large as possible in the two cases. A new design[110], having a time-dependent bias voltage was then proposed in which the harvested work and power could be increased by some orders of magnitude at low vibration frequencies, and by a factor of up to 5 at high frequencies. A more recently reported[111] oxidized liquid metal-droplet based energy harvester converts acoustic energy into electrical energy by modulating an electrical double layer that results from deformation of the oxidized droplet. As usual, a gallium-based alloy was chosen for such applications due to its high electrical conductivity and unlimited deformability. The proposed energy harvester consisted of top and bottom electrodes which were covered with a dielectric layer, with the gallium-based droplet placed between them. When an external bias-voltage and acoustic wave were applied, the contact-area between the droplet and the electrodes changed and thus led to variations in the capacitance in the electrical double layer and to the generation of an electrical current. This energy harvester exhibited a maximum output current of 41.2nA for an applied acoustic wave of 30Hz.

Another early design[112] involved a liquid-metal droplet-based tubular electrostatic energy harvester in which the droplet acted as a tribo-electrification material. Double-helix electrodes were placed around a tube so as to obtain cyclic transferred charges which multiplied the output frequency and markedly improved the efficiency. Such a tube was intended to be bound to a user's wrist and thus harvest energy from the wearer's movements. Irregular arm-swinging produced a voltage of up to 3.52V.

A micro thermomechanically pyro-electric generator has been shown[113] to be an alternative to thermoelectric generators for harvesting ambient heat-energy. In these devices, a thermal mass oscillates between hot and cold sides due to the bistability of its mechanical mounting. This generates a time-dependent thermal gradient in pyro-electric material which is located in between. The natural frequency is a major factor which governs the power output of the device, and this in turn depends upon the thermal contact resistance which exists at the contacting regions of thermal mass on the hot and cold sides. In order to reduce the thermal contact resistance, galinstan droplets are positioned at the contacting surfaces. The positioning is achieved via the selective deposition of galinstan onto a laser-micromachined silicon substrate. By incorporating such arrays, the frequency of the device can be increased by at least 50%.

There are several relevant popular deposition techniques, which use various methods to deposit liquid metals selectively. Embedded structures of liquid metal can be created by using stencils to pattern the metal before encapsulation. In order be able to remove the mask neatly, the alloy can be cooled to below its freezing point to solidify it, followed by casting an elastomer around it. Repetition of this process permits several layers of independent liquid metal to be embedded within the same elastomer for use as a capacitive sensor.

In another stencil-based process, a copper stencil is placed on partially-cured elastomer and an excess of liquid metal is poured onto the stencil and rolled flat using a roller. The latter also removes excess alloy. The stencil is then removed and the pattern of liquid alloy remains on the substrate before an elastomer layer is applied and cured. Stencils work well because the underlying elastomer can conform to it and thus form a good seal while, because of low adhesion, the stencil can be easily removed. The smallest features which can be made in this way are about 200µm, with a minimum spacing of 100µm. The stencil technique is thus relatively simple and reliable, but is suitable only for producing relatively large features of simple geometry. The liquid surface following deposition is also often rough and non-uniform in thickness.

Reconfigurable Antennae

Reconfigurable radio-frequency components are invaluable for their ability to change key parameters such as the operating frequency and gain. Normal reconfigurable components have to include switches or solid-state devices because the solid metallic elements of the circuit themselves cannot be altered. Their replacement by liquid metal permits reconfigurability. Antennae are an essential part of any wireless device, and a wearable antenna can serve both as a means of communication or as an energy-harvester. Such antennae, when used as a component of a wearable sensor network, have to possess qualities such as flexibility, stretchability and robustness. They may be sewn into fabrics, encapsulated in polymers, printed onto flexible laminates or injected into microchannels. The liquid nature of the moving parts of the antenna avoids major disadvantages of mechanically-actuated reconfigurable antennas, such as mechanical failure due to fatigue, creep or wear. Displacement of liquid metals can also be more easy, than in the case of solid metals, by applying microfluidic techniques such as continuous-flow pumping and electrowetting. Closely-related to electrowetting is the electrically-actuated movement of liquid metals. The spreading of EGaIn on a copper substrate in 0.5mol/l NaOH solution, using various direct-current polarities, has been investigated[114]. This showed that, when the copper substrate was connected to the cathode, wetting could be largely attributed to reduction of the oxidized surface of the substrate. Under the opposite polarity, spreading could be attributed to the tension gradient existing along the EGaIn/solution interface. The latter induced a Marangoni flow, even though the EGaIn–copper system was non-wetting here. In general, controlling the interface tension is an effective means for manipulating the flow, position and shape of fluids at sub-mm length-scales, because interface tension is here the predominant force. Most of the available methods for controlling interface tension are of limited use for controlling liquid metals however, due to the high interface tensions which are involved. Electrical means of control are nevertheless easier to miniaturize than are mechanical ones, in spite of various problems. Electrowetting on a dielectric requires kilovolt potentials, and electrocapillarity can produce only rather small changes in interface tension. Continuous electrowetting meanwhile is limited to manipulating slugs of liquid metal in capillaries. Gallium-based liquid metal alloys can best be controlled via electrochemical surface reactions[115]. Electrochemical methods can control the withdrawal of EGaIn from microfluidic channels because applying a reductive potential to the metal removes the oxide, when in the presence of an electrolyte, and induces capillary behavior. The process can be switched on and off repeatably by applying a voltage. The applied current is the key factor (figures 13 to 16) which governs the withdrawal velocity[116].

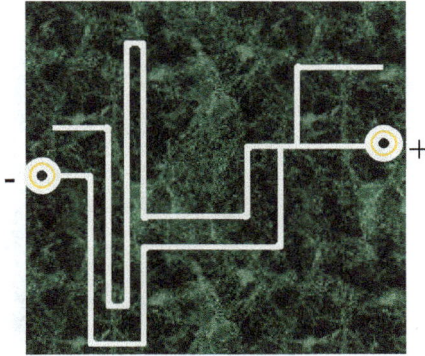

Figure 13. Schematic of the shape-reconfiguration of a liquid-metal microstructures by recapillarity. A microfluidic channel of arbitrary shape is filled with the liquid, mechanically stabilized by its oxide skin and 10mM NaF is added as the working electrolyte, with copper electrodes being attached to the inlet and outlet.

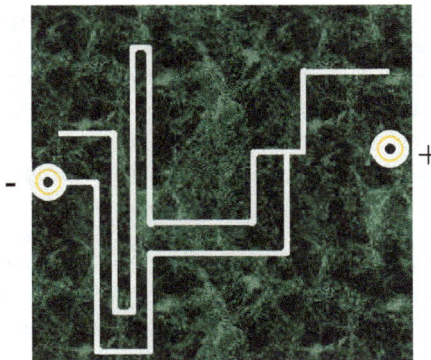

Figure 14. Application of some 40μA of reductive current induces withdrawal of the liquid

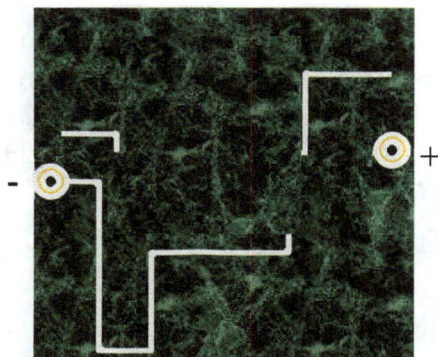

Figure 15. The electrodes control the withdrawal path, and the
process can be halted at any time by turning the voltage off

Experimental results suggest that the current is required in order to reduce the oxide at the leading interface of the metal, as well as the oxide which is sandwiched between the wall of the microchannel and the bulk liquid metal. It has been shown[117] that gallium-based liquid-metal alloy can be controlled electrically without any need for an external power supply. Liquid metal can be used as an anode to drive a complementary oxygen reduction reaction that results in the spontaneous growth of hydrophilic gallium oxide on the liquid metal surface and induces flow of the liquid metal into a channel. The extent and duration of the movement can be controlled throughout, and the process and induced flow are reversible and repeatable. A reconfigurable liquid-metal antenna, on an open 2-dimensional surface, consisted of EGaIn plus an electrolyte which could electrochemically modify the surface tension of the EGaIn and cause it to flow and contract on the open surface. By altering the shape of the EGaIn in this way, its properties could theoretically be varied over a wide range. A study of the geometries which could be obtained in this way showed that, the larger the electrolyte volume, the lower were the resonant frequency and radiation efficiency. When the electrolyte volume was 80mm x 50mm x 0.5mm, the operating frequency could be continuously varied from 0.71 to 4.03GHz, for a tuning ratio of 5.6:1, and the maximum gain was 3.48dB[118].

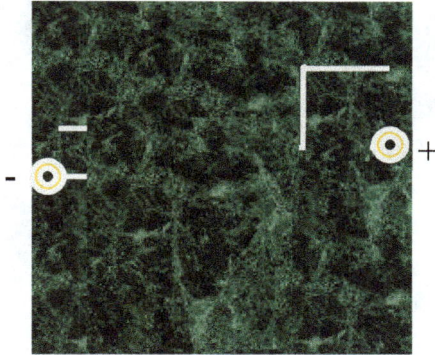

*Figure 16. Only that metal in the electrical path
withdraws when the voltage is turned on*

A crossed-dipole antenna which exhibited frequency and polarization variability was controlled[119] via the electrochemical actuation of liquid metal. Multidirectional displacement of the metal permitted frequency and polarization reconfiguration without any need for mechanical pumps or other devices. The dipole arms consisted of liquid metal volumes which could be shortened or lengthened within capillaries by applying direct current to each arm. Changing the lengths of the dipole arms generated two independently-tuned linearly-polarized resonances between 0.8 and 3GHz, and a polarization that could be switched from linear to circular between 0.89 and 1.63GHz.

Pressure-actuated liquid metal devices were demonstrated[120] for reconfigurable tunable dipole antennae, for use at GHz frequencies, which offered switchable shielding with 35dB attenuation, about 30dB polarizer attenuation and about 40° diffraction from a linear grating. These devices were also very flexible as they employed Ga-21.5-10.0%Sn alloy in a sealed system with an acidic vapour environment, together with non-alloying corrosion-resistant carbon-inks as electrical connectors[121].

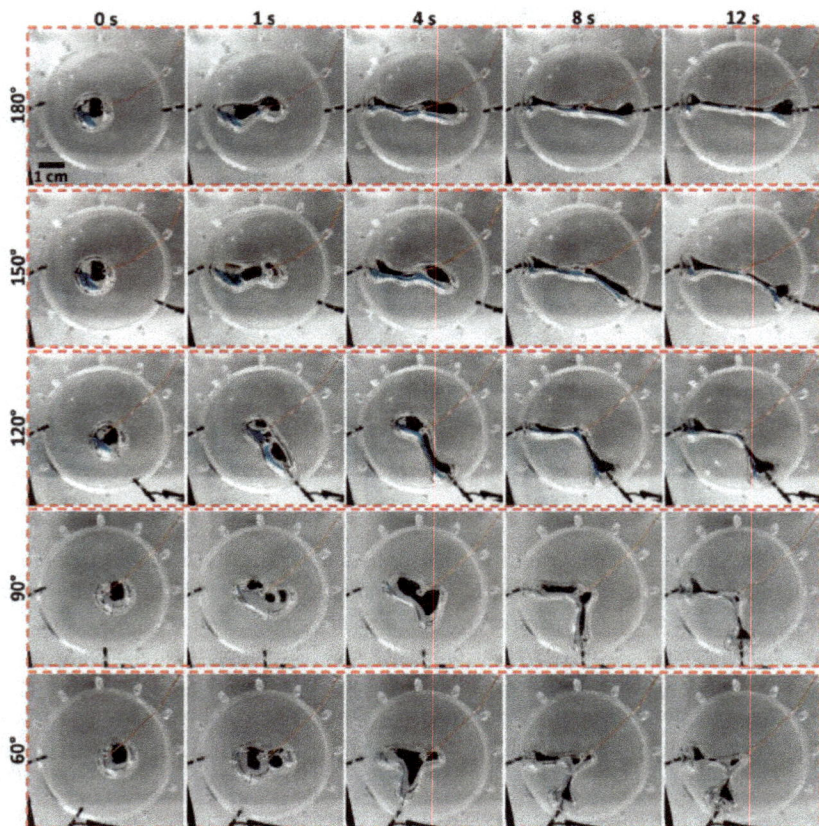

Figure 17. Automatic modification of the morphology of an EGaIn droplet into chosen patterns, having protrusion angles of 180, 150, 120, 90 and 60 ° (top to bottom), within 12s of activation of the control system. Reproduced under creative commons licence from: Automatic Morphology Control of Liquid Metal using a Combined Electrochemical and Feedback Control Approach, M.Li, H.M.C.M.Anver, Y.Zhang , S.Y.Tang, W.Li, Micromachines, 2019, 10, 209

Although various methods have been tried for the control of the morphology of liquid metals, there remains a paucity of methods that can ensure the precise morphological control of a free-standing liquid metal droplet which is not mechanically confined. Electrochemical manipulation of liquid metals is relatively easy, but there are few techniques that are open to precise automatic control. Investigations[122] have recently been made of the use of an electrochemical technique, combined with feedback control, to control automatically and precisely the morphology of a free-standing liquid metal droplet in sodium hydroxide solution. When a fairly small voltage is applied to a liquid-metal droplet, a resultant electrochemical oxidation of the surface can markedly decrease the interfacial tension. Although the oxide normally acts as a barrier to flow, the metal can nevertheless flow in the presence of a chemical base which competes with electrochemical deposition by dissolving the oxide. The morphology of the liquid metal can be controlled by using electrochemical techniques and a feedback control system. For example, the morphology of EGaIn droplets contained between laser-patterned acrylic sheets could be modified at will. The lower (50mm) plate consisted of a 6mm-deep chamber containing 0.4M NaOH solution and an EGaIn droplet while the upper plate allowed electrodes to pass through 12 small (1mm-diameter) holes arranged in a circle. An initial direct current from side-wall electrodes at 5V was used to ensure that the droplet could flow smoothly from an anode, at the centre of the chamber, to cathodes at the side-wall. When a lower voltage was applied, the droplet could not reach the cathodes; instead flattening and ceasing to move. That concentration of NaOH also allowed the EGaIn droplet to move smoothly and without much morphological distortion. A thick oxide layer formed on the droplet at lower NaOH concentrations, and again prevented the droplet from moving towards the cathodes. At higher concentrations, unwanted electrolysis occurred at the electrodes. The control system served to monitor the current passing between the droplet and one cathode, to compare it with the current value detected between the droplet and the other cathode and to determine the pulse-width of the modulation control signal. Under the combined influences of electrochemistry and feedback-control, the initial aim was to move the 1ml droplet from the centre of the chamber and towards the two cathodes so as to form 2 protrusions. The EGaIn droplet could finally be re-configured automatically and precisely into patterns having 2 main protrusions which intersected at 5 different angles: 180, 150, 120, 90 and 60°. Formation and elongation of liquid-metal protrusions towards the cathodes occurred, where the elongation was due to non-uniform oxidation of the metal on areas facing the cathode; a process which could locally reduce its interfacial tension. This could then generate a Marangoni flow which was directed towards areas possessing a higher interfacial tension; the centre of the chamber. This then dragged the liquid metal protrusions towards the cathode. The configurations (figure 17) of the EGaIn droplet were relatively stable, and could maintain their morphology while the control system was active. The droplet could not form stable protrusions if its volume was less than 300μl.

Also of interest is the question of just how small a microfluidic channel can be filled with gallium-based liquid metal. Due to the high surface tension of the liquid metal, it might be anticipated that the possible dimensions of microfluidic channels are somewhat limited. In order to determine the minimum possible channel dimensions, thick (5, 10, 25μm) polydimethylsiloxane-based microfluidic channels, having widths ranging from 20 to 3μm, were prepared[123]. The liquid metal was successfully introduced into channels with a width as small as 3μm and a thickness of 25μm; giving a cross-section of 75μm^2.

Continuous electrowetting is an effective mechanism for controlling reconfigurable radio-frequency devices by electrically inducing motion in a liquid-metal slug, but precise control of the slug position within fluidic channels is problematic. Precise positioning of liquid-metal slugs was possible using continuous electrowetting together with channels that were designed to minimize the liquid-metal surface energy at various locations. This approach exploited the high surface tension of the liquid metal so as to control its resting point to within sub-mm accuracy. The electrowetting actuation and fluidic channel design could be optimized so as to create reconfigurable radio-frequency devices. A reconfigurable slot antenna, produced by using these techniques, achieved a 15.2% tunable frequency bandwidth.

Altering the electrochemical potential at the surface of liquid metal in an electrolyte can change the interface tension by more than 2 orders-of-magnitude (about 500mN/m to almost zero); a change which is both rapid and reversibly. The required potential, of less than 1V, is also very reasonable. As just mentioned, the change in tension is due mainly to electrochemical changes in the surfactant-like oxide layer. Removing this oxide increases the interface tension, and the process is eminently reversible. The effects of the oxide are not always negative: whereas most liquids form beads in order to minimize surface energy, the presence of surface oxide on these liquid metals makes it possible to pattern them into useful shapes by, for example, fluidic injection and 3-dimensional printing.

Many different electrolytes can be used to remove the oxide, and the phenomenon occurs independently of the nature of the substrate. It has been argued[124] however that immersing the gallium alloy in a fluid which is also electrically conducting can diminish the electromagnetic effectiveness. Acidified siloxanes have been essayed which remove or prevent oxide formation; with the siloxane oil incorporating hydrochloric or hydrobromic acid. This oil is electrically insulating, chemically stable, removes etching by-products and water and permits electrowetting. Some remaining problems can be resolved by using so-called EWOD (electrowetting on dielectric) devices and a careful choice of electrode material, dielectric fluid and overall architecture[125]. A radio-frequency device, in the form of an electromagnetic polarizer, could be activated at will due to

EWOD and provided an average signal attenuation of 12.91dB in the on-state and 1.46dB in the off-state, between 8 and 9.2GHz, with a switching-speed of about 12ms. As well as being used for the reconfiguring of antennae, the electrical control of surface tension can also be used for the pumping of discrete droplets of liquid metal for cooling purposes; especially the removal of hot-spots from a solid surface. The use of liquid metals, rather than water or air, markedly increases the heat-transfer rate, while electrowetting is an efficient low-power low-voltage means for pumping liquids at the micro-scale. Early calculations[126] indicated that a more than 2 orders-of-magnitude increase in heat transfer rate was possible by using liquid metals at a flow velocity of more than 10cm/s, driven by a potential of just 2V.

It is also possible to manipulate these liquid metals by using magnetic fields, even though gallium-indium alloys are not ferromagnetic. As a demonstration, a so-called magnetic liquid-metal marble was described[127] which permitted the on-demand wireless control of a droplet for electrical switching purposes. The marble was created by coating the oxidized surface of the liquid metal with ferromagnetic iron particles having diameters of 45μm or 45nm. This led to non-wetting behaviour, as shown by measuring the static and sliding contact-angles against surfaces such as glass, Teflon and polydimethylsiloxane. In the case of Teflon-coated glass, the greatest (169.0°) contact angle was observed for 45nm-diameter particle-coated liquid-metal marbles, while the lowest (17.2°) sliding angle was observed for 45μm-diameter marbles, respectively. A minimum magnetic field of 150gauss was required in order to move 45μm-diameter iron-coated liquid-metal marbles. The presence of hydrochloric acid vapour increased the lyophobicity, leading to a sliding angle of 9.4°, and reduced the minimum magnetic flux density from 150 to 107gauss. In related work[128], the surface of liquid-metal droplets was electroplated with a CoNiMnP layer or was coated with iron particles by rolling the droplet on a bed of such particles. When placed on absorbant paper, the minimum magnetic flux density which could cause the movement of roughly 8μl-sized iron-particle coated droplets was 50gauss. As compared with the CoNiMnP-electroplated droplets, the iron-particle coated droplets could be controlled because the iron particles were uniformly distributed over the droplet surface. At a maximum applied magnetic flux density of about 1600gauss, the CoNiMnP layer broke down, resulting in fragmentation of the droplet. In the case of the iron particles, they were detached from the liquid metal surface and re-coated when the magnetic field was removed.

Returning for a moment to the surfactant oxide problem, a method which was based upon light-scattering and spectrophotometry was used[129] to determine the effect of organic surfactants upon the size and yield of the gallium-indium eutectic nanodroplets which were formed in organic solvents by ultrasonication. The process was used to deduce the role played by head-group chemistry and polar/apolar interactions of aliphatic surfactant systems in nanodroplet formation. Ethanol was the most effective solvent medium for the formation and stabilization of EGaIn nanodroplets. In the case of thiol-based surfactants in ethanol, the yield of nanodroplets increased as a function of the number of carbon atoms in the aliphatic component so that, in the case of the most effective surfactant (octadecanethiol), the nanodroplet yield was increased by some 370% over that of plain ethanol. Low overall efficiency of the reaction, together with an incompatibility of surfactant-stabilized EGaIn nanodroplets in non-polar organic solvents, indicated that the stabilization mechanism was different to the self-assembled monolayer process of nanoparticle formation. It has also been found[130] that surfactant-free sessile gallium-indium eutectic nanodroplets can be formed by exploiting the weak adherence which exists between the liquid metal and its ultra-thin oxide layer. Nanoscale liquid metal droplets could form, during the van der Waals exfoliation process involved in transferring the ultra-thin oxide layer onto a substrate, due to dynamic wetting failure. As an example, EGaIn nanodroplets could be prepared on the surface of transferred ultra-thin Ga_2O_3 films. Nearly all of the nanodroplets were isolated, randomly distributed and had a radius of less than 100nm.

In an early example[131] of a liquid-metal antenna, a 2.4GHz unbalanced loop antenna was created which could be stretched in several directions simultaneously. It involved the incorporation of room-temperature liquid metal alloy into microstructured channels in an elastic material. This prototype exhibited a stretchability of up to 40% along two orthogonal directions, together with foldability and twistability. Its radiation efficiency was greater than 80%. A mechanically flexible planar inverted cone antenna, intended for ultra-wideband use, could be folded and stretched significantly without suffering any permanent damage or loss of electrical continuity[132]. It was created by injecting room-temperature liquid metal alloy into micro-structured channels in an elastic dielectric. The latter, together with the metal, could be bent to a very small radius, twisted and stretched in any direction. The antenna had a return loss which was better than 10dB in the 3 to 11GHz range. Stretching by up to 40% was possible while maintaining electrical continuity. A later wide-band pattern-reconfigurable cone-antenna design[133] used liquid-metal reflectors. The layout consisted of a metal cone as the main radiator, and 8 poles as reflectors for beam switching. It had a wide impedance bandwidth. Beam-steering could be achieved by changing the states of the poles. A prototype, with 4 radiation modes, could perform 21 types of beam steering over a 360° sweep. A fractional bandwidth of 45.5% covered the range from 1.7 to 2.7 GHz.

Stage 1 (L_1,L_1)

Stage 2 (L_1,L_2)

Stage 3 (L_2,L_2)

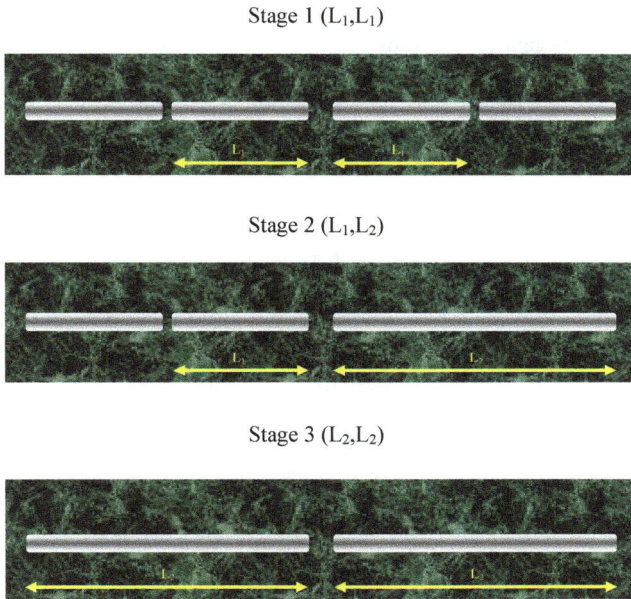

Figure 18. Frequency-shifting antenna at 3 different stages: L_1 is 12.6mm. Stage 1 consists of 4 isolated segments of liquid metal embedded in polydimethylsiloxane, with the 2 innermost segments defining the antenna. Application of pressure to the right-hand side of the antenna then causes two of the sections to merge into the new geometry of stage 2. Merging of the outermost with the innermost segments then produces stage 3.

Patch Antennae

A shape-shifting antenna was described[134] which changed its electrical length and therefore frequency when subjected to pressure. It was composed of liquid gallium-indium eutectic which had been injected into microfluidic channels containing rows of posts that separated adjacent parts of the metal (figure 18).

The initial shape of the antenna was defined mechanically by a thin oxide skin which formed on the liquid metal. Breaking this skin permitted the joining of separate portions of the metal. This rapidly changed the length and consequently the frequency of the antenna. A later flexible microstrip patch antenna was based[135] upon a multi-layer construction which again consisted of liquid gallium-indium eutectic (EGaIn) encased in an elastomer with serpentine channels that exploited the rheological properties of the liquid metal. Injection of the metal into microfluidic channels defined the shape of the liquid, which was again stabilized mechanically by the thin oxide skin on its surface. A more recent design[136] features a continually tunable microstrip patch antenna formed from EGaIn liquid contained within microfluidic channels in a polydimethylsiloxane substrate. This antenna can reconfigure its operating frequency in a continuous manner. The device consists of a microstrip patch which is excited via aperture coupling, and frequency tuning is arranged by altering the amount of metal within the channel so as to vary the electrical length of the antenna. The antenna is predicted to provide a frequency tuning-range of 3.21 to 10.12GHz, with a maximum real gain of 6.9dB and a total efficiency of 82%. The existence of a rapidly-formed oxide skin on the surface of EGaIn causes the latter to stick to surfaces and to limit the ability to reconfigure antenna shape. Water has been shown[137] to aid interfacial slip between EGaIn and other surfaces, thus allowing the metal to flow smoothly through capillaries and over surfaces without sticking. The presence of water also changes the chemical composition of the oxide and weakens it. This is not sufficient however to allow the metal to flow freely in microchannels without the surface layer. The slip layer offered new opportunities to control liquid-metal slugs in microchannels.

A square-patch frequency-selective surface made up of liquid metal embedded in a flexible silicone elastomer[138] was able to conform closely to doubly-curved surfaces. This facilitated electromagnetic interference reduction in the X-band frequency-range when wrapped over cylindrical or spherical radomes and dipole-like antennae.

An antenna has been constructed[139] from liquid metal and copper tape, attached to polyethylene terephthalate film. Polarization was controlled by the position of pressure-controlled metal, contained in 4 triangular cavities made from elastic dielectric material (Ecoflex) and polyethylene terephthalate film. This design enhanced the radiation efficiency while offering bendability. By using this arrangement, a reconfigurable aperture coupled patch antenna could be constructed which operated at 2.45GHz, with multi-polarization possibilities. Three kinds of polarization, including left-hand circular polarization, right-hand circular polarization and linear polarization were demonstrated, with a radiation efficiency greater than 90%[140].

A micro-strip patch array with 4 rectangular patch elements and a three-dimensional coaxial feed network, including power dividers and vertical transitions, has been embedded[141] in a 3-dimensional printed acrylic device, with the internal cavities being filled with gallium-based liquid metal alloy. Simulation and measurements of a 6GHz array showed that the latter produced a matched response and moderate gain at the design frequency. This construction permitted the integration of many radiating elements, and their feeds, into a single monolithic acrylic structure and thus avoided the need to produce separate printed-circuit-board antennae and feeds.

A parasitic circular patch antenna, fed at the center with a ring slot, has been developed[142] which exhibits continuously tunable linear polarization, with gallium-indium liquid being used to adjust the polarisation angle. The center resonant frequency of the antenna is 5.19GHz, with a −10dB impedance bandwidth of 0.24GHz. The rotational symmetry of the antenna helps to maintain the impedance and radiation pattern at the various polarization angles. The metal is contained in a 1mm x 1mm channel which is etched into a polymethylmethacrylate cylinder. The polarization angle is then varied by controlling the location of a short slug of EGaIn.

A novel wideband frequency-reconfigurable patch antenna with switchable slots, based upon liquid-metal manipulation in a 3-dimensional printed microfluidic channel, has been proposed[143]. Frequency-reconfiguration here relies upon switching the slots via the continuous movement of liquid metal over the channel that covers the narrow slots. One version comprised 3 pairs of composite slots, with the microfluidic channel bonded on top of the metal layer. By tuning and switching the liquid metal-loaded slot in the microfluidic channel a frequency tuning bandwidth of about 70% was possible, with a 2% instantaneous bandwidth. There was no appreciable change in the radiation pattern.

A recently proposed liquid-metal reconfigurable microstrip patch antenna features[144] an edge-truncated patch which is designed to produce circular polarization and incorporates 2 microfluidic channels for reconfiguration. The polarization of the antenna is switched to linear polarization, from circular polarization, when liquid metal is injected into the microfluidic channels. The latter are formed in polydimethylsiloxane, and the liquid metal is EGaIn. The measured 10dB impedance bandwidth is 1.579 to 1.619GHz and 1.563 to 1.587GHz for circular polarization and linear polarization, respectively. In the circular polarization case, the axial ratio at 1.595GHz was 2.267dB. In the linear polarization case, the axial ratio at 1.585GHz was 12.335dB.

A frequency-tunable microstrip patch antenna was described[145] in which the electrical length of the loading slot, located beneath the resonating patch, was altered by adding and removing liquid metal from fluidic channels along the ground plane. The slot reactively loaded the antenna, and altered the resonant frequency as its length was changed. The frequency could be tuned from 1.85 to 2.07GHz, corresponding to an 11.2% tuning bandwidth. A fully dielectric packaged liquid-metal patch antenna was based[146] upon 2 hollows embedded in a dielectric box, with a rectangular cavity on the upper part acting as the radiating element and a non-planar cavity having a central groove under the radiating element acting as the ground plane. The overall structure was prepared by means of 3-dimensional printing. The radiating element and ground plane were fully metallized by filling with liquid metal, with no need for solid metals. This antenna operated at 5.2GHz, and achieved satisfactory impedance-matching; with good agreement being found between simulations and measurements at the required frequency.

A flexible antenna, embedded in polydimethylsiloxane and integrated with an artificial magnetic conductor plane, was described[147] in which the conducting elements were incorporated by using liquid gallium-indium eutectic; apart from the ground plane. A dual-band capability, which was based upon that of a patch antenna modified with slots and slits, permitted its operation in the 2.45 and 5.8GHz bands. Integration of the antenna and artificial magnetic conductor plane, and the use of liquid metal embedded in polydimethylsiloxane, produced a highly flexible antenna, with the possibility of adding tunable characteristics to the structure. Measurements of the antenna indicated bandwidths of 80 and 460MHz in the lower band and upper band, respectively, and between 4 and 5dB of gain.

A mechanically controllable variable-frequency antenna was developed[148] for wireless communication within the frequency range of 0.5 to 3GHz. The variability was made possible by adjusting the floating ground plane dimensions under a meander patch by using galinstan having an electrical conductivity of 3.46×10^6S/m at 20C. The galinstan was gradually injected into channels by using a syringe.

A reconfigurable design involved the 3-dimensional printing of polymethylmethacrylate to form a double patch antenna[149]. Channels in the assembly contained galinstan immersed in 2% sodium hydroxide solution. When a voltage was applied across the channel, it changed the surface tension at the galinstan/NaOH interface, thus permitting it to flow across the channel. The resonant frequency could be tuned by using an external voltage to re-drain and re-fill the channel. It was predicted that the center frequency of the operational range would shift from 14.20GHz, when no voltage was applied, to 15.10GHz when the actuation voltage was applied across the channel.

A frequency-reconfigurable microstrip patch antenna has been created[150] by injecting galinstan into a square reservoir within a silicone elastomer. An aperture-coupled feed was used in order to increase the strain by separating the fixed-feed element from the stretched radiating element. The patch antenna could be stretched to strains of up to at least 300%, and the electrical length varied with the degree of extension, permitting frequency-tuning from 1.3 to 3GHz. A maximum radiation efficiency of 80% was measured.

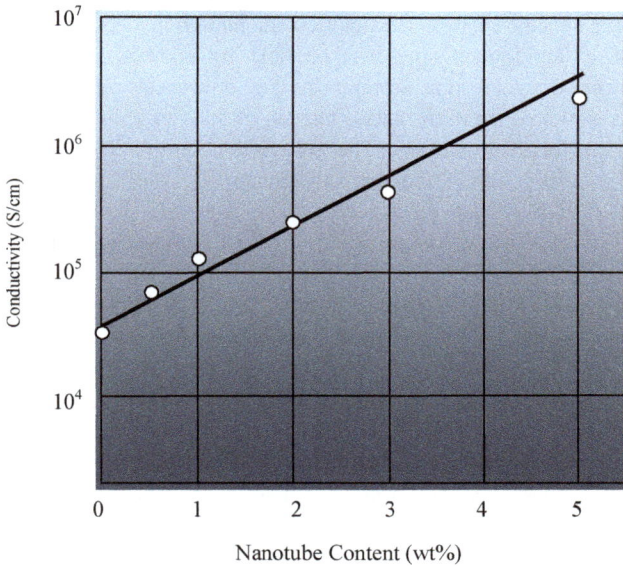

Figure 19. Conductivity of EGaIn nanocomposites as a function of nanotube content

A fluidic patch antenna, operating in the S-band (4GHz), comprised[151] a nanocomposite which was made from liquid gallium-indium eutectic blended with single-wall carbon nanotubes and operated at a frequency of 4GHz. This nanocomposite was enclosed in polydimethylsiloxane polymer by using ultra-violet light-assisted direct writing[152]. There was an increase in the electrical conductivity and reflection coefficient with carbon-nanotube concentration (figure 19).

Liquid metal has been used[153] to tune the operational frequency of a monopole antenna by adjusting its length. The centre frequency could be continuously tuned over two octaves: 2.0 to 9.5GHz, while maintaining a gain of about 5.2dB.

A broadband monopole antenna for stretchable electronics has more recently[154] been constructed using a galinstan conductor for the radiating element and ethylene-butylene-styrene as the antenna substrate. Simulated data were in good agreement with measured reflection coefficients, which indicated broadband operating characteristics between 1.4 and 6GHz. Continuous electrowetting was shown[155] to be an effective actuating mechanism for reconfigurable radio-frequency devices containing liquid metals. Electrowetting is a method for electrically inducing motion in a liquid-metal slug. Precise positioning of a liquid-metal slug was possible by combining electrowetting with channels that were designed so as to minimize the liquid-metal surface energy at certain locations. This exploited the high surface tension of the liquid metal so as to control its location to within less than one mm. The low-friction surroundings of the liquid-metal slug can lead to difficulties if the final position of the slug in the channel is to be controlled to a high degree of accuracy. One way of limiting unintended slug displacement is to increase the surface area of the liquid metal relative to its volume. This in turn increases the friction which is experienced by the slug. Experiments showed that, for a constant channel width of 2mm, a channel height of 600μm can lead to the slug being moved through up to 5mm after the actuation signal ceases. When the height was reduced to 200μm, the movement was reduced to 1 or 2mm; an uncertainty which is unacceptable. On the other hand, a reduction in height also makes it more difficult to control the slug.

The high surface tension of galinstan can be used to obtain more precision in slug-positioning. One method is to furnish the slug with localized points, within the channel, where it can minimize its surface energy. The surface tension of oxide-free galinstan in a nitrogen environment has been measured to be about 535mN/m, and it is probably slightly lower when the galinstan is immersed in an electrolytic solution.

A high surface tension makes the liquid metal sensitive to variations in channel geometry because they directly affect the surface area of the slug and hence its free surface energy. Channels were prepared so as to take advantage of galinstan's high surface tension: instead of having a uniform cross-section, the channel included a series of interlocking circular chambers. These acted as local minima in the surface energy of the liquid metal. When the continuous electrowetting actuation signal was removed, the strong internal cohesion of the slug forced it into the nearest energy minimum. The channel was thus effectively divided into a finite number of discrete overlapping locations at which the slug could rest following actuation. This increased the accuracy of its resting point.

The continuous electrowetting and channel design together created reconfigurable radio-frequency devices. A reconfigurable slot antenna, based upon these methods, exhibited a 15.2% tunable frequency bandwidth. Additional solutions were found for minimizing the tendency of galinstan to form a surface oxide layer which wetted the channel walls and thus inhibited motion. One method of dealing with the oxide has been to attack it with sodium hydroxide. In the case of monopole antennae, the electrolyte changes the effective length of the antenna, thus causing it to have a lower return loss, at lower frequencies, than that predicted by the metal height alone[156]. A more recent tunable liquid-metal antenna design[157] used a reconfigurable stub which was made from galinstan. This 5GHz antenna offered more than 10dB of analogue gain-tuning, from -5.90 to 4.43dB. By exploiting continuous electrowetting, a 60Hz signal with an amplitude of 1V and a 75% duty cycle could drive liquid metal continuously along a channel so as to tune the stub length and antenna gain. Zero external power was required in order to maintain the position of the slug.

A previous reversible resonant frequency tunable antenna[158], based upon liquid-metal control, consisted of a co-planar waveguide-fed monopole stub printed onto a copper-clad substrate, plus a tunnel-shaped microfluidic channel which was linked to the printed metal. The gallium-based liquid metal was added to, or removed from, the channel by using air pressure. Control of the liquid metal made it possible to change the physical length of the antenna reversibly, and at will. The gallium-based metal had been treated with hydrochloric acid in order to remove the oxide layer and its attendant wetting and sticking effects. Removal of the oxide layer permitted repeated reversible tuning between 4.9 and 1.1GHz. Phosphonic acid has also been used[159] to control the interfacial chemistry. This again leads to zero adhesion, between the gallium surface oxide and substrates, and no alloying with metal electrodes.

Another approach is to render the channel surfaces oxide-repellent by spray-coating with commercially available products. Careful examination showed[160] that these coatings consisted of silica nanoparticles, linked by silicones, which exhibited 2 length-scales of roughness, and surface roughness is known to impede adhesion. It was confirmed however that both hydrophobic and hydrophilic rough surfaces prevent oxide adhesion. The coatings permitted reversible actuation, via sub-mm channels, so as to form reconfigurable antennae in the GHz range without having to use corrosive acids or alkalies to remove oxides. The coatings also permitted open-surface patterning with conductive traces of liquid metal. It was therefore possible to actuate liquid metals in air without leaving metal or oxide residues on the surface.

The EGaIn residue, because of its rapid oxidation, limits multiple movements of EGaIn in reconfigurable radio-frequency components, and so some attention has been focused[161] on the use of surfactants, carrier liquids and microchannel coatings in order to minimize EGaIn fragmentation and residues on polydimethylsiloxane-based microfluidic channels during repeated actuation of an EGaIn plug. This led, by using a combination of carrier liquid and microchannel coatings, to a minimization of the tendency of EGaIn to leave residues on polydimethylsiloxane microfluidic channels. A microstrip transmission line switch was developed, for reconfigurable radio-frequency use, by using an EGaIn plug. It was switched on with less than 4dB, and off with a loss of less than 18dB.

Complex mechanically reconfigurable microfluidic devices have been made from EGaIn in styrene-ethylene-butylene-styrene thermoplastic elastomer, with gecko-based adhesive bonding[162]. Selective filling of the microfluidic structures was carried out via controlled movement of the EGaIn/air interface by using hydrophobic valves. The latter took account of the critical Laplace pressure of EGaIn so as to fill multiple branches and arrays of isolated elements. Bow-tie, folded and multilayer dipoles, and microstrip patch antennae, were prepared which had a thickness of less than $600\mu m$[163]. They could be made to conform to complex surfaces, and could be stretched or bent in order to change electromagnetic properties such as the center frequency, bandwidth and impedance-matching. As an example, a mechanically reconfigurable folded dipole was created which had 55% frequency tuning, with a 20 to 40% bandwidth for various frequencies.

Liquid-metal reconfigurable antennas are commonly flushed with an electrolyte in order to remove the nm-thick oxide of EGaIn which sticks to the walls of the microchannels. This EGaIn residue prevents repeatable actuation, but the presence of the electrolyte causes other problems due to its often corrosive and electrically conductive aspects. A technique for aiding the reversible infusion and withdrawal of EGaIn into acrylic micro-channels and wide planar cavities involved[164] coating the surfaces with a silica-particle based superhydrophobic coating. The coating prevented adhesion of the liquid metal to the acrylic, and allowed it to be reconfigured repeatedly without requiring the usual flushing.

A half-mode substrate-integrated waveguide antenna has been proposed[165] in which the resonant frequency was to be switched by injecting EGaIn into a single vertical fluidic hole, the position and size of the latter being chosen so as to maximize the tuning-ratio and peak gain. The resonant frequency could in fact be switched from 2.33 to 3.35GHz, by injecting EGaIn into the vertical fluidic hole, while keeping the peak gain above 6dB.

Expanding upon this, a method was developed[166] for producing multiple resonances in a radio-frequency planar structure without requiring any additional circuitry or passive elements. The structure was based upon substrate integrated waveguide technology and involved a mechanism by which 16 distinct resonant frequencies (table 3) between 2.45 and 3.05GHz could be produced by perturbing a fundamental frequency. This structure involved the use of 4, rather than 1, vertical fluidic holes and the injection of EGaIn into those channels so as to produce various resonant frequencies. Having a vertical channel either empty, or filled with metal, therefore led to 16 frequencies by exploiting all of the possible combinations of the binary states of each of 4 vertical channels. In related work[167], a frequency-switchable complementary split-ring resonator-loaded quarter-mode substrate-integrated waveguide band-pass filter has been designed in which the frequency switching involved the use of microfluidic channels and EGaIn.

Table 3. Resonant frequencies and Q-factors of the 16 metal-injected states of a substrate-integrated waveguide resonator

State	Channel 1	Channel 2	Channel 3	Channel 4	Frequency (GHz)	Q-Factor
1	0	0	0	0	2.45	24.22
2	0	0	0	1	2.69	22.52
3	0	0	1	0	2.56	25.4
4	0	0	1	1	2.85	22.75
5	0	1	0	0	2.48	21.4
6	0	1	0	1	2.77	20.73
7	0	1	1	0	2.69	20.31
8	0	1	1	1	2.92	19.91
9	1	0	0	0	2.50	19.5
10	1	0	0	1	2.95	19.43
11	1	0	1	0	2.59	19.31
12	1	0	1	1	2.90	19.24
13	1	1	0	0	2.53	19.18
14	1	1	0	1	2.85	19.15
15	1	1	1	0	2.74	19.1
16	1	1	1	1	3.05	19.08

The channels were made from polydimethylsiloxane. The device was designed to have 2 states. Before injection of liquid metal, the center frequency and fractional band-widths were 2.205GHz and 6.80%, respectively. Following injection, the center frequency shifted from 2.205 to 2.56GHz and, although the coupling coefficient was essentially unchanged, the fractional band-width changed from 6.8 to 9.38% as the complementary split-ring resonator shape changed and the external quality factor decreased. Following the removal of liquid metal, the measured values were similar to those found before liquid metal was injected.

A first attempt was made some years ago[168] to switch fluidically through via-posts in order to obtain a switchable quarter-mode substrate-integrated waveguide antenna. The switching method was based upon filling or emptying a non-plated via-hole using liquid metal. Such an antenna was designed so as to work initially at about 3.2GHz. Connecting a fluidically switchable via-post at the corner of the quarter-mode substrate-integrated waveguide antenna shifted the operating frequency upwards, giving a switching-range of 3.2 to 4.7GHz.

With the object of being able to expand previous research on liquid-metal linear arrays into progress on dynamically driven 2-dimensional liquid-metal arrays, simulations were performed in order to support the concept that a multi-dimensional array would provide more control over the directions of the main beam and of the nulls; thus making the resultant array useful for the spatial tracking of signals and also for minimizing interference. A prototype model of such a 2-dimensional antenna array was tested[169] in the form of a dual-frequency global positioning satellite antenna, and the results were comparable to those predicted by the simulation. The liquid-metal construction of the antenna then made it possible to change the parasitic elements of the 2-dimensional array by using a syringe or pump. The spatial adaptability of this antenna also permitted the use of a single digitizer chain and thus offered a possible alternative to anti-jamming global position satellite antennae.

A first 3-dimensional printed reconfigurable helical antenna, based upon microfluidics and liquid metal alloy, was described[170]. Three-dimensional printing techniques were used to form a helical microfluidic channel which was then filled with EGaIn. The gain of the antenna was determined by the number of turns in the helix, and this in turn was controlled by the volume of liquid metal. A better than 4dB gain-increase at around 5GHz was found when the number of turns in the helix was increased from 2 to 8. This corresponded to just a 0.2ml change in the volume of liquid metal.

A reconfigurable antenna, with mechanically actuated liquid metal held in a helical channel, was proposed[171] for the frequency band of 1.575GHz. The antenna consisted of a flexible polymer tube, in the form of a helix, which was filled and emptied using a computer-controlled peristaltic micropump. The antenna could generate 4 types of beam: linearly polarized semi-doughnut and axial and elliptically and circularly polarized axial. The corresponding maximum gains were 1.1, 5.9, 7.6 and 8.5dB.

A polarization-reconfigurable conical helical antenna has been based[172] upon liquid metal by using a truncated structure, variable pitch angle, matching stub and a mechanical autorotation device. The gain of the antenna attained more than 8.5dB in the 1525 to

1660.5MHz band, and the 3dB axial ratio bandwidth reached 520MHz. The polarization mode could be switched between right-hand and left-hand circular polarization[173].

An early microfluidics-based approach[174] to developing a reconfigurable circularly-polarized transmitting array comprised double-layer nested split ring slots formed as microfluidic channels that could be filled with liquid metal. Split regions in the slots were obtained by injecting liquid metal into the channels. Beam-steering was ensured by provoking rotational phase-shifting via manipulation of liquid metal in the slots. X-band unit-cell prototypes were created on glass substrates having a patterned metal film, and slot channels were formed from polydimethylsiloxane. In one variant[175], only the inner ring had a split region, and the latter was created by leaving a gap in the channel which was filled with injected liquid metal while the outer channel was filled with liquid metal. The transmission phase of the unit cell was again adjusted by causing rotational phase-shifting via manipulation of the liquid metal in the channels. Movement of the liquid metal, together with the split around the ring, provided a 360° linear phase-shift range in the transmitted field through the unit cell. A circularly polarized unit cell was designed[176] so as to operate at 8.8GHz, and satisfied the required phase-shifting conditions of the element rotation method. The behaviour of unit-cell prototypes confirmed the waveguide simulator method between 8 and 10GHz. In another self phase-shift dipole array having beam-scanning capability, the basis of the design was a 2-element dipole array in which the arms of 2 dipoles were made of EGaIn and the capillaries could be shortened or lengthened by applying direct-current voltages[177]. As a result of the length changes, the self-impedance characteristics of the dipoles were changed and thus imparted beam-scanning properties. So as to increase directivity, 2 movable EGaIn parasitic elements which acted as director and reflector were present in 2 separate microfluidic channels. Electromagnetic band-gap structures, placed between the 2 dipoles, reduced any potential mutual coupling. Predicted results indicated that the radiation patterns could cover scan-angles ranging from -21° to 21°, with the gains being stable at 7 to 8dB and the isolation being better than 20dB.

Liquid-based tunable periodic structures are capable of acting as highly frequency-selective surfaces or spatial phase-shifters, and of providing phase-shifts ranging from 0 to 360°. The practical devices consist of multi-layered periodic structures which are made up of non-resonant unit cells. The tuning mechanism is based upon integrating small movable liquid-metal droplets with the unit cells of the periodic structure. By moving the liquid-metal droplets over small distances within the unit cell, the frequency response of the structure can be continuously varied. As an example[178] such a fluidically-tunable frequency-selective surface, offering a fifth-order band-pass response, was designed and its tuning performance was investigated for various incidence angles and polarizations of an incident electromagnetic wave. Electronically tunable equivalents having the same structure were also examined under short-duration high-power excitation conditions. The electronically tunable devices exhibited extremely non-linear responses whereas the fluidically-tunable structures did not involve any non-linear devices and their response was therefore expected to remain linear under short-duration high-power excitation conditions.

The intermodulation distortion which is produced by passive and tunable liquid-metal antennae has been investigated[179] by using 4 types of monopole: passive copper, varactor-tuned copper, passive liquid-metal and tunable liquid-metal. The linearity was assessed by using a two-tone distortion test: 2 fundamental tones, close in frequency, produced third-order intermodulation tones at frequencies above and below the original tones, separated by the frequency difference. The passive liquid-metal monopole possessed a comparable linearity to that of the passive copper monopole, while the linearity of the electrochemically controlled capillarity-tuned liquid-metal monopole was at least 40dB better than that of the active varactor monopole. The reconfigurable liquid-metal antenna could also handle higher powers (31dB) before failure than could the active varactor-tuned antenna (24dB).

A beam-reconfigurable superstrate antenna consisted[180] of a periodic flexible surface which comprised an array of metal-filled microchannels within a polymer. The beam could be reconfigured by stretching the surface along one axis, and the radiation pattern here depended upon the percentage elongation. It could also be reconfigured by re-shaping the surface to be concave or convex, and the pattern here depended upon the surface state. Such elongation or re-shaping could split the broadside beam into 2 beams which were up to ±55° or ±58° off-broadside, respectively. This design avoided the need to incorporate switches or varactors into the device.

Liquid metals are generally injected manually, into the inlet of a microchannel, by using a syringe but such injection is possible only if air already in the channels can escape through outlets. It is essentially the same problem encountered during conventional casting. The considerable pressure, proportional to the yield stress and inversely proportional to the channel dimension, which is required in order to inject liquid metals can also cause leaks or a failure in integrity of the channels during injection. It is found that complex 3-dimensional antennae can be created[181] by vacuum-filling gallium-based liquid metals into 3-dimensional printed cavities at room temperature. The cavities are produced by co-printing a sacrificial-wax together with acrylic resin. Removal of the wax then leaves cavities, which can be as small as 500μm, within monolithic acrylic. The entire structure is then put under vacuum, with the air escaping via a covering layer of liquid metal. Restoration of atmospheric pressure then pushes the metal into the cavities, leaving planar and curved conductive geometries with no pockets of trapped air. Experimentation with a simpler model[182] of this filling process suggested that the elastomeric channel walls absorb residual air which is displaced by the metal. Thus even without the aid of this vacuum treatment, the metal may be able to fill branched structures and culs-de-sac as small as a few microns within seconds and without any need for outlets. It was also possible to fill completely snaking microchannels which were up to meters in length. The smallest achievable feature size is also reported to be 250μm. Certain refinements of the method may be required. In one variant, for example, an initial excess of the liquid metal is poured onto the polydimethylsiloxane mould. When the vacuum is now applied, trapped air which comes from the mould may remain in the liquid alloy. Following de-gassing, the liquid metal layer has to be rendered free from holes by agitating the mould under vacuum. Restoration of atmospheric pressure then fills the mould completely and the excess material is removed. The liquid is then frozen, before removal and encapsulation in a new polydimethylsiloxane block. A porous plastic film with very small pores can help to fill a mould quickly when placed above the liquid metal. The liquid metal cannot pass through the sub-micron pores, whereas air can easily do so and is rapidly removed. Vacuum-assisted infiltration thus offers a higher filling-rate than does needle injection, but may have to be repeated in order to remove every vestige of trapped air.

In a two-dimensional design[183], liquid metal alloy was embedded in a structural composite panel in the form of micro-channels. The metal could be pumped in and out of the panel, where the antenna geometry was a derivative of the zig-zag log-periodic dipole wire type. The bandwidth and frequency response could be controlled by altering the amount of alloy in the meandering channel layout.

A K-band reflective waveguide switch was based[184] upon using gallium-indium eutectic, in which the switching mechanism involved inserting or removing metallic posts (one or two rows, 3 three posts per row) in the broad wall of the waveguide. Insertion losses which were as low as 0.1dB were found in the ON state, and an isolation of better than 30dB was found in the OFF state, when operating at 20GHz.

A wide-band antenna operating in the microwave regime was formed[185] by encasing liquid metal in a highly stretchable elastomer. It was arranged in the form of a two-arm Archimedean spiral so that it could be inflated into the form of a dome by using micro-blowers. It operated in the frequency-range of 6.9 to 13.8GHz with good circular polarization, and the possibility of changing the shape made it more directional while the passing band remained wide.

A fluidic antenna, made from polydimethylsiloxane filled with galinstan, comprised[186] a feed-line and 2 square shapes; all formed from liquid metal. The gallium-based liquid metal adhered to the channel surface, due to the usual viscous layer of gallium oxide. The liquid was treated with hydrochloric acid solution in order to remove the oxide, leading to easy movement of the metal in the channel as it was then non-wetting. The physical length of the liquid metal slug in the feed-line was controlled by air pressure, allowing the resonant frequency to be tuned between 2.2 and 9.3GHz.

An electrically actuated reconfigurable liquid-metal dipole antenna was described[187] in which, via electrocapillary actuation, a 5V direct-current signal could put galinstan into 5 discrete states having various polarizations and null directions. Local surface-energy wells, built into the polyimide encasing the fluid, led to metastable locking of the galinstan, thus removing the need to impose a continuous direct-current bias voltage in order to maintain each state.

A frequency-tunable half-wavelength dipole antenna was created[188] from an array of electrically actuated liquid-metal pixels. The resonant frequency of a half-wavelength dipole antenna depends upon the electrical length of the dipole arms. The use of liquid-metal pixels to alter the dipole's length results in discrete changes in the frequency of the antenna. One particular device demonstrated frequency-reconfiguration by switching between resonances at 2.51, 2.12, 1.85 and 1.68GHz. The liquid metal was actuated by changing its surface tension via continuous electrowetting. That is, the metal was immersed in 1M sodium hydroxide solution in order to form an electrical double layer at the metal/NaOH interface. A voltage which acted upon the double layer then created a surface-tension imbalance on the liquid metal, resulting in pressure actuation of a 4mm x 4mm liquid-metal so-called pixel. The liquid-metal pixelated antenna was based upon a 64mm-long baseline planar copper dipole, mounted on a 0.787mm-thick substrate. A liquid-metal pixelated antenna replaced a section of both dipole arms with a 1 x 4 pixel array, in which the walls of the array were made of polyimide. The upper side of the array was covered in polystyrene and the underside was covered in polydimethylsiloxane. Adjacent pixels on the upper side were interconnected using stainless-steel connectors that were embedded between the pixel walls, while the pixel array was connected to the copper section of the antenna by a soldered stainless-steel wire. This use of stainless is quite usual because the gallium-based liquid metals amalgamate with copper. The interface between the copper and the pixel array was covered with a watertight polymer. In order to turn a pixel on, an amount of liquid metal was electrically caused to move from a reservoir buried below the antenna. In order to turn a pixel off, the liquid metal was caused to retreat to the buried reservoir. A 1.2V square-wave with a +1V direct-current offset was applied to electrodes and thus brought the liquid metal from the reservoir. The liquid metal could be caused to return to the reservoir simply by inverting the polarity of the voltage which was applied to the electrodes. An alternative 3mm x 3mm pixel required actuation by a 30Hz 4V square-wave with a +1V direct-current offset, and a larger actuation force than the 1.2V actuation voltage used for the 4mm x 4mm version. The effect of a larger force, acting on a smaller body of metal, was to increase the actuation speed greatly. In the most recent example of this technique, liquid-metal pixels were used[189] to produce a frequency-tunable dipole antenna which could be configured so as to achieve horizontal or vertical polarisation. Each dipole consisted of 14 pixels, with of them being electrically actuated using galinstan so as to provide various antenna states.

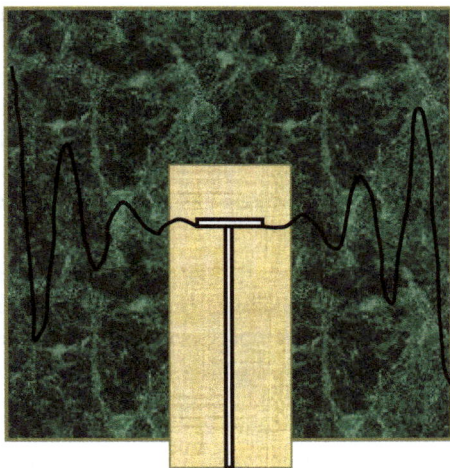

Figure 20. Vascular liquid-metal antenna with sinusoidal networks

A so-called vascular antenna was formed[190] from a thin-wire dipole made up of 2 curvilinear networks (figure 20) that were embedded in a composite panel and connected to a conductive vertical interconnect structure and a parallel-strip transmission line. Pressure-driven control of liquid metal within the network permitted physical reconfiguration of the meandering dipole so as to tune the matched impedance bandwidth of the fundamental dipole mode by symmetrically lengthening the conductive pathways which formed the dipole. The flow of liquid metal through the channels could also be used[191] to limit the temperature of the composite in which the antenna was embedded. The liquid metal antenna was highly reconfigurable with regard to the electromagnetic behaviour, and the thermal energy which was generated during high-power operation could be dissipated by circulating or cyclically replacing the liquid metal. The device was capable of out-performing copper-based antennae in high-power situations[192].

Another reconfigurable structurally-embedded vascular antenna was created[193] within flat epoxy/quartz-fiber composite panels and was based upon the transport of liquid metal (less than 2% by volume) within embedded microchannels. Patterned microchannels were fabricated via the fused deposition printing of sacrificial catalyzed polylactic acid, and the later removal of the latter. When the channels were progressively filled with liquid metal and electromagnetically connected, the resonant frequency could be tuned over a large frequency range that depended upon the resultant shape of the liquid-metal layout. The microvascular panels suffered a slight decrease in tensile strength due to the presence of the microchannels. Such reconfigurable liquid-metal structurally-embedded vascular antennae were anticipated[194] to incorporate a plasmonic nanoparticle-based method for measuring internal temperature gradients, and embedded micro-cantilever carbon-nanotube based sensors and thus accelerate the development of structural concepts by providing air-flow measurements and structural feedback during the flight-testing of aircraft.

A so-called origami tree antenna was introduced[195] which permitted the integration of multiple 3-dimensional antennae, with EGaIn being used to switch between the antennae. An origami structure termed a zipper tube, together with a Voronoi topology, was used as a scaffold. This facilitated mechanical tuning of the radiation pattern while minimizing the space requirement. The tree was fabricated by 3-dimensional printing, and consisted of 2 zigzag/helical antennae which permitted 2-band (3 and 5GHz) use and linear/circular polarization; plus giving various radiation patterns upon tree compression. Another fanciful design was that of a tunable bow-tie antenna[196]. Stereolithography permitted fabrication of the structure, which was then metallized using liquid metal alloy so as to permit folding without breaking. The proposed bow-tie antenna had a frequency-tuning range of 896 to 992MHz and offered bandwidth reconfigurability. A pattern-switchable origami antenna had previously been made[197] from paper by using inkjet-printing technology. That antenna could be switched between loop and dipole modes by folding and unfolding the paper, respectively. Liquid gallium-indium eutectic was used to avoid cracks forming in the conductive ink when the paper was folded.

A reconfigurable frequency-selective surface was designed[198] for use in bi-state transmittance/reflectance applications. The liquid metal, EGaIn, was again encased in a flexible dielectric substrate. The device could be switched from dual-polarized all-pass, to polarization-selective bandpass/bandstop behavior, by injecting EGaIn into the microchannels. The asymmetrical structure exhibited band-pass behavior under vertical polarization, but exhibited band-stop behaviour under orthogonal polarization. It functioned well, under various angles of incidence, for both polarizations. The structure exhibited narrow-band absorption at 5.92GHz during vertical polarization, but this shifted to 15.09GHz under horizontal polarization. The design also exhibited[199] a so-called miniaturization performance during the band-pass response at 1.37GHz; corresponding to unit-cell dimensions of $0.059\lambda_0$ x $0.059\lambda_0$. In a further development[200] of the idea, the frequency-selective surface consisted of 2 layers of periodic meandering patterns, encased within the opposite sides of a dielectric, with the layers arranged in orthogonal polarization and with independent control of the reconfigurability. This permitted switching, by injecting liquid metal into the top and bottom microchannels, between 4 different working states: dual-polarized all-pass, single-polarized low-pass, single-polarized band-pass and dual-polarized band-pass. The design again exhibited miniaturization performance during the dual-polarized band-pass response at 1.58GHz; corresponding to unit cell dimensions of $0.052\lambda_0$ x $0.052\lambda_0$.

A galinstan band-pass filter with good flexibility was prepared[201] by using an improved soft lithography process. The liquid metal was enclosed within the microfluidic channels of a polydimethylsiloxane substrate, and the filter could be bent, twisted and stretched in any direction. A second-order open-loop resonator filter having a 2.4GHz center frequency and 15% fractional bandwidth was created in order to test the design. Experimental data showed that the frequency-response of the filter was essentially unaffected by bending or twisting, although the center frequency shifted slightly downward when stretched. A new tuning method for coaxial-cavity resonator-based band-pass filters has been based[202] upon pneumatically-controlled galinstan which constituted the inner conductor of a capacitively-loaded coaxial cavity resonator. By changing the height (i.e. volume) of the galinstan in the inner conductor, ultra-wide (more than an octave) and low-loss radio-frequency tuning could be achieved without requiring the small capacitive gaps which usually limit the radio-frequency performance of a coaxial band-pass filter. As an example, a coaxial-cavity resonator which permitted frequency-tuning between 2.9 and 9.9GHz (a 3.4:1 tuning range) and had an unloaded quality-factor of between 120 and 625 was prepared. The radio-frequency performance of a 2-pole band-pass filter having a tunable frequency of between 3.4 and 7.5GHz (2.2:1 tuning range), an intrinsically switched-off state, an insertion loss of less than 1dB and a return loss of less than 10.5dB was described.

A mm-wave steerable/reconfigurable phased-array antenna, with EGaIn liquid metal switches, has been produced[203] which operates at 26.2GHz and has a scan range of ±58°;

giving a ±40° improvement in scan-range over conventional dipole phased-array antenna at a side-lobe level of at least 8.8dB. This beam-scanning method offers continuous beam-steering over a much wider scan-angle range than is possible by using conventional techniques.

An 8-element 1-dimensional Ka-band focal plane array which was capable of beam scanning was described[204] which was based upon microfluidic principles. The array was placed at the back surface of an 8cm-diameter extended hemispherical Rexolite dielectric lens and consisted of interconnected microfluidic reservoirs and channels which had been constructed[205] by bonding together polydimethylsiloxane and liquid crystal polymer substrates. The antenna part of the array was a small volume (2.5µl) of liquid metal contained within low-loss Fluorinert solution. The array beam was then scanned by moving the liquid-metal antenna among the reservoirs by using a low-loss dielectric solution and an external pump. The proximity-coupled feed network of the array was passive and was designed so as to be able to handle the positional changes of the liquid-metal antenna element. The array operated with a 7° half-power beam-width, a greater than 21dB gain and provided a ±30° beam-scanning range[206]. This microfluidic-based beam-scanning technique also functioned without any need for radio-frequency switches.

Another liquid-metal scheme[207] for continuously tuning the frequency of a reconfigurable antenna involves EGaIn, contained in microfluidic channels and controlled by an impressed voltage. Electrochemical deposition or removal of a surface oxide on the EGaIn markedly lowers or increases its interfacial tension, and this changes the length of EGaIn in the channel without pumping. In one case, small direct-current voltages were used[208] to tune an antenna continuously and reversibly between 0.66 and 3.4GHz, resulting in a 5:1 tuning range while the radiation efficiency varied from 41 to 70% over the tuning-range.

Via a coaxial cable, 5 EGaIn monopolar antennae are connected so as to form a bendable array. This could change both the size of the liquid antenna, but also increase the reconfigurable frequency range. The antenna is predicted to provide frequency tuning from 1.3 to 11.3GHz, and the adjustable-range ratio is about 8.7:1. The maximum real gain is 4.09dB and the maximum total efficiency is 82%. The above technique had previously been applied to a reconfigurable Yagi-Uda antenna[209]. Control of the length of liquid EGaIn in a capillary had also been carried out via the electrochemical deposition or removal of the surface oxide. This again markedly lowered or increased its surface tension and caused the liquid to move up or down the capillary. A 3-element array in that case consisted of components the lengths of which could be changed by moving the liquid metal. As a result, the array directors could be transformed into reflectors, and tuned to various frequencies. It was predicted that good impedance-matching was possible from 0.85 to 0.95GHz, with a gain of over 5dB. The utilization coefficient of the aperture field reached 70%.

In a similar scheme[210], EGaIn was contained in a highly stretchable elastomer in the form of a cylindrical helical antenna. The interfacial tension of the EGaIn was again increased or reduced by exploiting the electrochemistry at the EGaIn surface by forming or destroying an oxide layer. The resultant capillary force was used to drive EGaIn flow in the microfluidic channels. The antenna was continuously adjusted by means of a low direct-current voltage. This controlled the length of the radiating unit and effected changes in the resonant frequency of the antenna. The antenna was expected to work at frequencies ranging from 0.75 to 1.25GHz, with a gain of 5dB.

A frequency-reconfigurable quasi-Yagi dipole antenna was designed[211] which exploited microfluidic technology by combining a metal-printed driven dipole element with 3 directors. In order to tune the resonant frequencies, microfluidic channels were integrated into the driven element and, in order to maintain a high gain for all of the tuned frequencies, microfluidic channels were also integrated into the directors. The length of the driven element, as well as directors, could thus be controlled by injecting liquid metal into the microfluidic channels. The amount of liquid metal which was injected into the channels was controlled by using programmable pneumatic micropumps. One version exhibited continuous tuning of the resonant frequencies from 1.8 to 2.4GHz, while the measured peak gain ranged from 8 to 8.5dB.

A reconfigurable 5-element Yagi-Uda monopole array used[212] pressure-driven liquid-metal elements. There was a single centrally-located driven element, and 6 parasitic elements in an array, which had their physical lengths varied by using liquid metal. This allowed the array to be tuned to various frequencies, as well as being able to vary the number of director elements, and change parasitic elements from reflectors to directors. Beam-steering along a single axis was possible, with 1.33 dB of gain at 2.4 and 3.87GHz. In another liquid-metal reconfigurable Yagi-Uda monopole, a 3-element array consisted[213] of driven and parasitic elements whose lengths could again be changed using pressure-driven liquid metal and which resulted in an array where directors could be transformed into reflectors, and where the elements could be tuned to various frequencies. The radiation patterns again demonstrated single-axis beam steering at frequencies from 1.95 to 4.11GHz.

A flexible frequency-reconfigurable coplanar waveguide-fed antenna consisting of 4 split-ring resonators having differing radii has been developed[214]. A negative mould was fabricated by ultra-violet lithography and polydimethylsiloxane was then used to cast a positive structure. Liquid metal was injected into the microfluidic channels of the polydimethylsiloxane. The pass state of the split-ring resonators could be changed by mechanical pressure, thus allowing the antenna to operate at frequencies ranging from 1 to 6GHz.

A microfluidic impedance tuner, fabricated on polydimethylsiloxane, has been combined[215] with a planar inverted-F antenna by using a simple double-stub in which the impedance was changed by tuning the stub length via the injection of liquid metal into the microfluidic channels using a piezoelectric micropump. The antenna initially operated at 900MHz, with an impedance matching of 50Ω. The impedance became mismatched when experimenters stood close to the antenna, and the mismatched impedance was again matched to 50Ω by injecting the liquid metal. At 900MHz, the return loss could be tuned from 4.69 to 18.4dB when an experimenter's hand was placed 1mm above the antenna. An earlier approach[216] to producing a miniaturized inverted-F antenna operating at 885MHz had involved a 3-dimensional flexible arrangement based upon liquid metal and additive printing methods. The sensitivity to bending was greatly reduced by this arrangement. Another frequency-tunable liquid-metal planar inverted-F antenna[217] was used to compensate for hand-effects on mobile phones. The upper arm of the antenna consisted of a galinstan-filled Teflon tube. The length of the filled portion determined the resonant frequency, and the tuning of the antenna could be adjusted automatically so as to recover from disturbances caused by hand-contact.

A glass dielectric resonator antenna has been described[218] which incorporates a liquid-metal polarizer that is capable of effecting polarization reconfiguration. The polarizer is formed from the usual gallium-indium-tin alloy, and the antenna is able to generate 3 different polarizations: -45° polarization, +45° polarization and y-axis polarization. The glass dielectric resonator antenna is designed to operate at 2.4GHz, with a wide effective bandwidth of 18.0% and a radiation efficiency of greater than 80%.

Methods of integrating liquid metal with microfabricated devices have been investigated[219], together with the corrosion of metal interconnects and liquid fill-factor effects due to stretching. A macro-scale monopole antenna was able to tune the frequency, via elongations of up to at least 40%. Account had to be taken of the fact that the increased length affected the electrical resistance of the metal and the resonant frequency of the antenna. A novel spiral-shaped device was made, using Parylene-C and dispersed liquid metal, which exhibited complete filling and promised elongations of up to 200%.

Galinstan wiring with a 3-dimensional coil has recently been developed[220] for preventing resistance changes during device deformation. Core-shell hydrogel microsprings had previously been prepared by using a double bevel-tip nozzle. Liquid metal was here instead injected into a microspring core which was composed of starch and a 3-dimensional coil of liquid metal developed. The resultant metallic wiring maintained a stable resistance, with less than 1.2% variation, during 100% tension and a stable impedance, with less than 0.2% variation, for alternating currents ranging from 1 to 100kHz.

A controllable-beamwidth impulse-radiating antenna has been designed[221] in which the usual solid reflector of the antenna was replaced with a wired reflector corresponding to the current flow on the reflector surface. The behaviour of the wired-reflector impulse-radiating antenna was essentially the same as that of the normal impulse-radiating antenna. Liquid metal was subsequently used to replace the solid wires, and a beamwidth-controllable wired-reflector impulse-radiating antenna resulted. The electrical size of the reflector could be expanded by up to about 400mm in the H-plane, and the beamwidth in that plane could be varied by up to 44% while parameters such as the input impedance remained essentially constant.

By using the soft silicone substrate concept, and combining it with the typical designs of conventional patch and ground-plane microstrip antennae, two forms of stretchable microstrip antennae have been produced: the meshed-microstrip antenna and the arched-microstrip antenna[222]. The former antenna exploits the initially wavy structures that result from patterning, while the latter exploits the deformed wavy structures that result from pre-straining. When compared with the equivalent solid-metal microstrip antenna, the radiation properties of the stretchable antennae are little changed. The resonant frequency decreases, with externally applied tensile strain along the feed direction, in the meshed-microstrip antenna but increases with increasing strain in the arched-microstrip antenna.

Two different designs for frequency-reconfigurable antennae, capable of continuous tuning, have been proposed recently[223]. In both cases, the radiator is an aperture-fed microstrip patch which is formed from liquid metal contained within a microfluidic channel structure in polydimethylsiloxane. In one design, the microfluidic channel structure has a meandering layout and incorporates rows of posts. It is predicted to provide a frequency-tuning range of some 118% (4.36GHz) between 1.51 and 5.87GHz, and experimental results for the fully-filled case indicate a resonance at 1.49GHz. Straight channels are used in the other design and a frequency-tuning range of about 77% (3.28GHz), from 2.62 to 5.90GHz, is predicted.

A centre-fed finite-length infinitesimally-thin dipole antenna, for use as a pressure-sensor in geotechnical applications, consists[224] of liquid gallium-indium eutectic in microfluidic channels housed in polydimethylsiloxane. It is stretchable and frequency-reconfigurable and can act as a pressure-sensor when the length of the polydimethylsiloxane changes under pressure. There is a linear relationship between the resonant frequency of the antenna and the applied pressure, as well as between the resonant frequency and the displacement.

A polarization-reconfigurable antipodal dipole antenna[225] switches between 2 orthogonal linear polarizations under the low-power electrical actuation of liquid metal. Unlike earlier electrically-actuated liquid-metal antennae, which required a continuous applied voltage and suffered reduced gain due to exposure to the lossy electrolyte needed for actuation, the present antenna uses metastable locking. That is, an applied voltage is required in order to change the operating state rather than maintain it. The compact design also reduces the required amount of electrolyte. The dipole has a maximum gain of between 1.8 and 2.3dB, in the presence of electrolyte, and between 2.4 and 3.0dB when the solution is absent. The resonant frequencies of the orthogonal polarizations are less than 5.3% apart, and all of the states have a greater than 15dB return loss at the 3GHz operating frequency.

Another frequency-tunable antenna design involved printing with an acrylic-based polymer, blanket-metallization with a thin copper layer and feeding with a 50Ω coplanar waveguide[226]. Frequency-tunability was achieved by injecting EGaIn into meandering and interconnected 3-dimensionally printed fluidic channels bonded with polydimethylsiloxane. The antenna operated in the frequency range of 2.6 to 8GHz.

A microfluidically-reconfigurable frequency-tunable liquid-metal monopole antenna was developed[227,228] which depended upon the continuous movement, using a micropump, of a liquid-metal volume over a capacitively-coupled microstrip line-feed network. In order to maximize capacitive coupling at the feed-point, the antenna was constructed by bonding microfluidic channel moulds of polydimethylsiloxane with a 25μm-thick liquid-crystal polymer substrate. The antenna could be used as a component of wideband frequency-tunable high-gain antenna apertures, without requiring additional micro-pumps, by exploiting meandering or interconnected microfluidic channels. A 4 x 1 microfluidically-controlled monopole array was able to provide a 2:1 frequency-tuning range from 2.5 to 5 GHz.

A pump-free method for controlling the length of liquid metal in a capillary, and thus changing the operating frequency of a monopole antenna, has been demonstrated. An applied direct-current voltage was used to control the surface tension of the liquid-metal filament and cause it to lengthen or contract; thereby varying the antenna's resonant length. A closed-loop feedback system was used[229] to keep track of the antenna operating-frequency and could adjust the applied voltage so as to shape the liquid metal into the desired form.

The reconfiguration of filled and empty microchannels has been used[230] to produce a switchable dual-band slot antenna in which the bands could be controlled independently. The reactive loading effect of galinstan bridges was used to shift the operating frequency of each slot antenna downwards. The use of 5 microchannels produced a 5-bit reconfigurable antenna, and any combination of frequencies between 1.8 and 3.1GHz (first band) and between 3.2 and 5.4GHz (second band) could be obtained. This was equivalent to a discrete tuning-ratio of about 1.7:1 for both bands, and an overall ratio of 3:1 (at 1.8 to 5.4GHz) for the antenna.

As a result of the reactive loading, the frequency of the antenna was reduced[231] and the antenna could be shrunk by a factor of 85%. By changing the configuration of filled and empty channels, each channel could be used as a switch. By using 2 pairs of microfluidic channels, 3 different frequency bands (2.4, 3.5, 5.8GHz) could be obtained; leading to a switching ratio of better than 2.5.

In some designs[232], the liquid metal is moved by pressure-driven air bubbles and guided within channels so as to reconfigure the length of the radiating aperture and feed line of a slot antenna. The gallium-based liquid is held in position by air bubbles and liquid motion is produced when a pressure differential exists. Unlike hydraulic or pneumatic control-methods, pressure-based control is reversible and repeatable provided that the liquid metal is enveloped by a thin layer of NaOH solution which lubricates, and removes oxidation. A slot antenna could achieve a 26% tunable bandwidth over a range of 1.42 to 1.84GHz. The peak gain ranged from 4.8dB in the lowest-frequency state to 4.1dB in the highest-frequency state.

Sensors

Some early success was had[233] in incorporating liquid metal as an interconnect within a super flexible sensor skin which could simultaneously monitor temperature and force. The sensing element was polydimethylsiloxane elastomer, mixed with 10% by weight of multi-walled carbon nanotubes. Galinstan was used to fill microchannels, as an electrical interconnect. This skin-like material could be bent sharply, with no fear of failure. The temperature coefficient of resistance attained 0.4%/C, and the resistance-change per unit applied force was about 0.3%/kg. A later passive temperature sensor was proposed[234] which used microfluidics and liquid metals to engineer temperature-related changes in the radar-echo of the sensor. Liquid metal could alter, as a function of temperature, the numbers of antenna elements which were activated along a linear array. The temperature could then be accurately deduced from the change in radar cross-section. The number of elements to be activated was controlled by their short-circuiting due to temperature-induced expansion of liquid metal within a contacting microfluidic channel. Simulations and measurements of the back-scattered power of such a temperature-reconfigurable array confirmed the sensitivity and temperature-range of the sensor. It offered[235] a tunable temperature-range of at least 20K and a resolution of 1.8dB per activated element, resulting in a temperature resolution of about 4K.

A self-contained large-area wireless strain sensor, which operated at about 1.5GHz, was later based[236] upon the concept of a multi-layer microfluidic stretchable radio-frequency device. The sensor was capable of detecting tensile strains of up to 15% remotely over large surfaces or movable parts … with no need for hard-wiring. Most of the sensor consisted of a mechanically reconfigurable and reversibly deformable patch antenna, with the latter in turn comprising 2 layers of liquid metal-filled microfluidic channels in a silicone elastomer. A simple radio-frequency transmitter consisting of miniaturized rigid integrated circuits could be assembled on a flexible printed circuit board and then linked to the antenna. Such an elastic patch antenna could survive repeated stretching, and still retain its electrical properties to some extent, before returning to its original state following removal of the stress. The electrical properties at the operating frequency were highly sensitive to mechanical strain. The antenna could therefore not only radiate and detect radio-frequency signals, at a range of 5m, but could also serve as a reversible large-area strain sensor. A normal force, and shear force, sensor was developed[237] which used galinstan as a piezoresistive material, encapsulated in polydimethylsiloxane. The use of liquid metal as a gauge material offered the advantage that it could detect large forces without breakage and that, because liquid-metal piezoresistors deformed together with the elastomeric surroundings, both normal and shear forces could be deduced from resistance-change data. Each sensor included a pair of symmetrical piezoresistors which were screen-printed onto a cavity in a polydimethylsiloxane substrate, at a tilt-angle of about 30°, so as to be sensitive to both normal and shear forces. A normal force would therefore compress both piezoresistors equally, while a shear force would shorten one piezoresistor but elongate the other. A flexible multilayer capacitive microfluidic normal-force sensor was also developed[238] which used liquid-metal filled microfluidic channels as capacitor plates and interconnecting conductors. It could be manufactured by using soft-lithography techniques, and consisted of multiple layers of polydimethylsiloxane microchannels, filled with galinstan and air-pockets, which modified the properties of the sensor. A single element, calibrated for normal forces ranging from 0 to 2.5N, could provide repeatable measurements of static uniaxial loads, and could monitor dynamic loading-unloading loads at between 0.4 and 4Hz. The initial sensor had a spatial resolution which was of the order of 0.5mm and functioned reliably when wrapped around surfaces having a curvature of 1.575/cm; and could potentially handle curvatures of up to 6.289/cm. The sensitivity and range of the sensor could be adjusted so as to, for example, exhibit a greater sensitivity at low loads. An artificial hair-cell sensor has been made[239] by encapsulating liquid metal in a polydimethylsiloxane substrate. Drawing upon the experience gleaned from the above flexible force sensor, which could detect normal and shear forces, the artificial hair cell sensor could detect 2-axis tactile forces via a standing artificial hair-shaft. Because the liquid-metal piezoresistors deformed with the elastomeric substrate, the normal and shear forces could be detected due to resistance changes in the piezoresistors. Each sensor consisted of a pair of symmetrical piezoresistors, which were again screen-printed onto a suspended polydimethylsiloxane membrane, with opposed orientations, so as to be sensitive to shear forces.

Considering more robust applications, the dipole liquid-metal antenna has been suggested[240] as a possible crack sensor when embedded in a concrete beam. The proposed antenna consisted of EGaIn, injected as usual into microfluidic channels within polydimethylsiloxane. Such fluidic dipole antennae were of course highly flexible, stretchable and reversibly deformable. Changes in length due to stretching of the elastomer channels also changed the resonant frequency. Increasing the length of the antenna, when it was not embedded in concrete, decreased the resonant frequency. The question was whether the antenna would behave in the same way if it were already embedded in concrete. In later work[241], concrete specimens were subjected to center–point loading tests while the resonant frequency of the embedded liquid antenna was measured simultaneously. Statistical analysis of the results showed that there was a significant relationship between the displacement of the concrete specimen and the resonant frequency of the embedded antenna (figure 21). This figure reveals a clear relationship between the displacement of a concrete beam, cured in 7 days, and the resultant resonant frequency of the antenna. The data can be described by: frequency (MHz) = 686.08 – 15.23 x displacement (mm).

Further work on such flexible piezoresistive shear and normal force sensors, based upon EGaIn, revealed[242] that there was a slight problem with regard to hysteresis in the normal force direction. The EGaIn could detect forces exerted on a flexible and stretchable substrate without the connection breaking, but had to be packaged in elastomeric protection. This however led to hysteresis, in the sensed resistance, with respect to changes in applied force. The results for static force, and for various rates of loading-force cycling, identified the relationship between the applied normal force and the hysteresis in the monitored signal. That is, the hysteresis was due mainly to the greater deformation and slower recovery-time of the elastomer which effectively governed the shape of the liquid metal wire used for the resistance measurements.

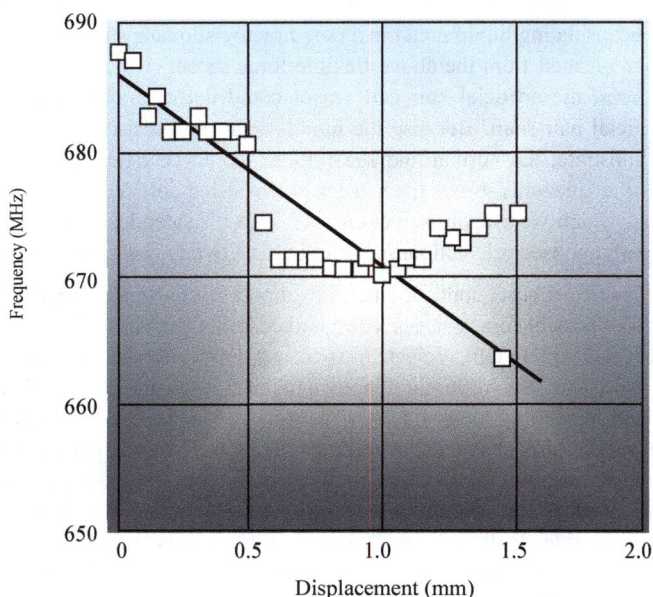

Figure 21. Resonant frequency versus displacement of antenna embedded in concrete

The above theme, of liquid metal and elastomer as core and shell materials respectively, was used [243]to develop a coaxial printing method for the preparation of stretchable conductive cable. When liquid EGaIn was embedded in an elastomer matrix under proper control, the resultant cable exhibited an acceptable mechanical performance, even when stretched by more than 350%. Even under compression, the cable could recover its original properties due to the high flowability of the liquid metal and the super-elasticity of the elastomer shell. The cable also exhibited high reliability, over many cycles, with regard to stretchability and conductivity. It was expected to be useful as a strain sensor for monitoring the motion, of a curved object, with regard to frequency and amplitude. Soft conductors can be created by containing liquid-metal nanoparticles between 2 elastomeric sheets. The particles initially form an electrically insulating composite, but these so-called soft circuit-boards can be written-on by using a hand-held stylus, and this sinters the particles into conductive paths due to the localized mechanical pressure applied to the elastomer sheets[244]. Antennae having tunable frequencies can then be prepared by sintering the nanoparticles into microchannels. A proposed liquid metal-based antenna for wearable applications consisted of a loop of silicon tubing, having a diameter of 1.5mm, into which the metal was injected[245]. The antenna's operational frequency was initially tuned to 868MHz, but it could be easily changed to 2.4GHz.

The development of the field of so-called soft robotics depends upon integrating the electronic components required for sensing, power-regulation and signal processing. Soft robots are designed to use their compliance to be able to interact safely with the environment and to move around it in ways that a rigid robot cannot. To more completely achieve this, the robot should be made of as many soft components as possible. A method was recently described[246] for the incorporation of micro-electronic sensors and integrated circuits into the elastomeric skin of a so-called soft robot. The thin stretchable skin contained the solid-state electronic devices required for the detection of orientation, pressure and temperature-level; all connected by thin-film copper wetted with EGaIn. An easy approach to the integration of highly flexible and stretchable microfluidic channels into textile-based substrates is to embroider micro-tubing directly onto a surface in a zig-zag or meandering pattern and then coat it with an irreversibly bonded elastomer[247]. This approach has been used to produce a stretchable drug-delivery system and a wearable wireless resonant displacement sensor which was capable of detecting strains ranging from 0 to 60%, with an average sensitivity of 45kHz/% strain, by filling the embroidered tubing with liquid metal, thus creating stretchable conductive microfluidics which exhibited a less than 0.4Ω resistance variation at maximum (100%) stretch. Such interconnects could moreover survive 1500 stretch and relax cycles to 30% strain, with a less than 0.1Ω change in resistance.

A completely soft hydraulic control valve has been described[248] which consisted of a 3-dimensional printed photopolymer body, with an electrorheological working-fluid and gallium-indium-tin liquid metal alloy as electrodes. This soft 3-dimensional printed electrorheological valve weighed less than 10g. Its pressure-holding capability was tested under unstrained conditions, as well as under cyclic activation and bending, twisting, stretching or indentation conditions. The maximum holding pressure of the valve was 264kPa when 5kV was applied across the electrodes, and that pressure deviated by less than 15% from the unstrained maximum holding pressure under all of the straining conditions, apart from indentation; where there was a 60% increase.

Printed electronics which comprise polymeric semiconductors offer a means for linking sensor-based circuits. An electrochemical pH-threshold indicator was proposed[249] which was based upon a printable hybrid electrode, with galinstan embedded in a conducting polymer matrix. This hybrid electrode exhibited a wide variation in open-circuit potential versus pH in an electrochemical cell. When connected to the gate of an electrochemical transistor this led to a sharp change in the drain current over a narrow range of pH values.

Direct ink writing of EGaIn has been used to create soft sensors in a stable manner; a process in which it is very important to maintain the flatness of the substrate, while the wettability of on the substrate by EGaIn greatly affects the print quality. The very different degrees of adhesion of liquid-metal inks, in various states of oxidation, to divers substrates can lead to inferior circuits and to consequent difficulties in preparing high-accuracy devices, or to low production-rates. Direct-writing and needle-injection are the most popular methods for patterning microstructured liquid metal alloys, but direct writing is complicated by the low viscosity and high surface tension of the metal. This causes the liquid to break up into droplets rather than remaining as a stream. It generally requires meanwhile a very high pressure in order to inject liquid metal. One solution[250] is to extrude liquid metal coaxially within an encapsulating fluid in order to preserve a continuous and stable stream. So-called fused deposition modelling 3-dimensional printing co-extrudes a liquid-metal core together with a shell made from thermoplastic elastomer. Such a system can produce conducting micro-wires having a diameter of some 25µm, together with an insulating shell of styrene-ethylene-butylene-styrene. This can be stretched to up to 4 times its original length without suffering any detectable mechanical or electrical loss. The inevitable mechanical damage which will eventually occur due to long-term solicitation nevertheless remains a problem. A magnetic healing method, based upon iron-doped gallium-indium conducting ink has recently been proposed[251]. This permitted remote self-healing by means of magnetic field. The presence of degradable polyvinylalcohol and adhesive fructose also permitted water-degradable and thermal transfer printing. A specimen light-emitting diode circuit could be structurally and functionally repaired following single- or multi-point damage, and the self-healing time of the latter damage was less than 10s. Due to the water-soluble polyvinylalcohol film, recycling involved just simple immersion in water and, by heating, an electric circuit on fructose could be efficiently transferred to another flexible substrate. Electronic devices possessing soluble or degradable characteristics are still in an early stage of development, due largely to a limited choice of suitable materials. Such so-called transient circuits could be made, as above, by combining liquid metals with water-soluble polyvinylalcohol. These transient circuits can exhibit considerable durability and a stable electrical behaviour during bending and twisting, while having short transience times due to the easy solubility of polyvinylalcohol substrates and the enormous flexibility of liquid-metal circuits. They are also eminently recyclable, with up to 96% of the liquid alloy being recovered. Surface patterning of liquid metals into complicated shapes is possible, offering the possibility of forming transient antennae for devices such as passive near-field communication tags[252].

The self-healing possibilities, also alluded to above, are of interest with regard to improving the capacity and durability of negative electrodes in lithium-ion batteries. Materials such as silicon, germanium and tin possess theoretical capacities which are greater than that of graphite, but tend to have a short cycle life due to fractures which are provoked by diffusion-induced stresses and by large volume changes which occur during electrochemical cycling. A higher capacity and improved durability can be achieved by using low-melting point alloys. In the case of gallium[253], a reversible solid-liquid transition occurs during lithium-insertion or removal. Cracks which form in the lithium-inserted solid state can thus be healed when the electrode returns to the liquid-gallium state following lithium removal. That is, the cracking failure-mode can be eliminated by using liquid-metal electrodes.

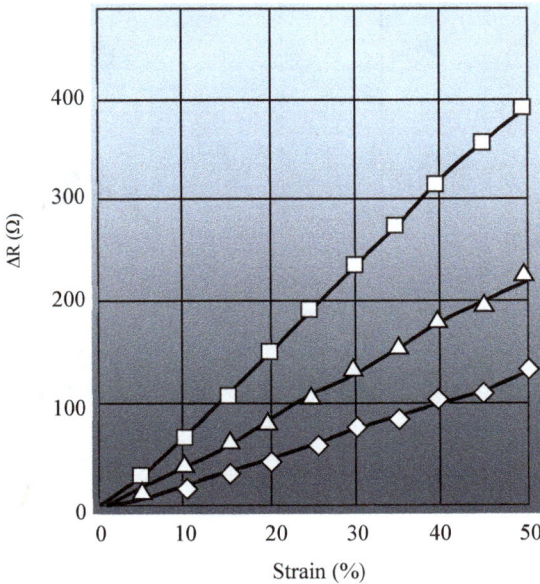

Figure 22. Variation of the resistance with strain, of sensors of various widths
Squares: width = 50μm, triangles: width = 70μm, diamonds: width = 90μm

Another method for the preparation of 3-dimensional microchannels at room temperature is to direct-write liquid metal as a sacrificial template[254]. The formation of the usually problematic surface oxide skin on the liquid metal is here exploited to stabilize the shape of the printed metal so as to form planar and non-planar structures. The usual fabrication methods limit microchannels to 2 dimensions, but the geometry of printed microchannels can here be varied from simple 2-dimensional networks to complex 3-dimensional structures without using lithography. The latter can be embedded in soft elastomeric or rigid thermosetting polymers. Acids, or electrochemical reduction, can be used to remove the oxide skin. This destabilizes the ink so that it drains from the encapsulant under capillary forces, thus resulting in almost complete recuperation of the ink at room temperature. The method produces monolithic structures without requiring bonding or other assembly. The removal of discrete portions of the metal leaves 3-dimensional features that can be used as antennae, interconnects or electrodes.

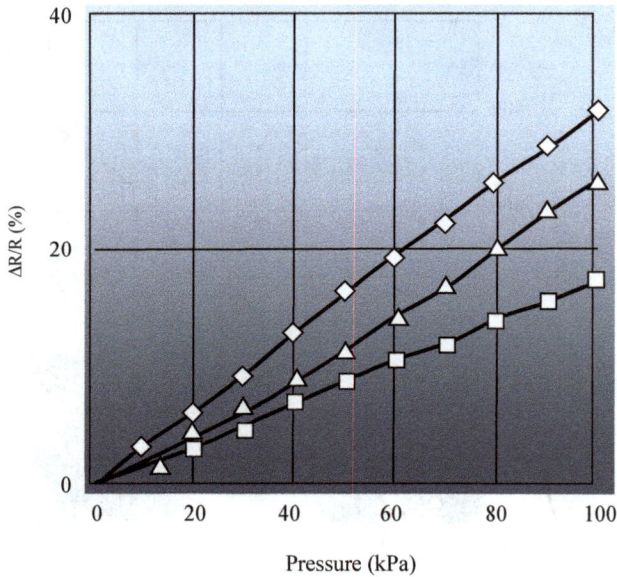

Figure 23. Normalized change in resistance as a function of applied pressure
Squares: width = 50μm, triangles: width = 70μm, diamonds: width = 90μm

The coating of inkjet-printed patterns of silver-nanoparticle ink, with a thin layer of EGaIn, can increase the electrical conductivity by 6 orders-of-magnitude as well as markedly improving the resistance to tensile straining[255]. The improvement is due to a room-temperature so-called sintering process, during which the liquid alloy bonds the circa 100nm-diameter silver nano-particles into a continuous conducting pattern. Ultra-thin, hydrographically-transferrable electronics could be produced by printing silver nano-particle plus EGaIn ink onto 5μm-thick tattoo paper. Such a printed circuit was sufficiently flexible to remain functional during deformation and could suffer strains of more than 80% with only slight electromechanical coupling. The thin-film circuits could be transferred to highly curved and non-developable 3-dimensional surfaces, or to human skin.

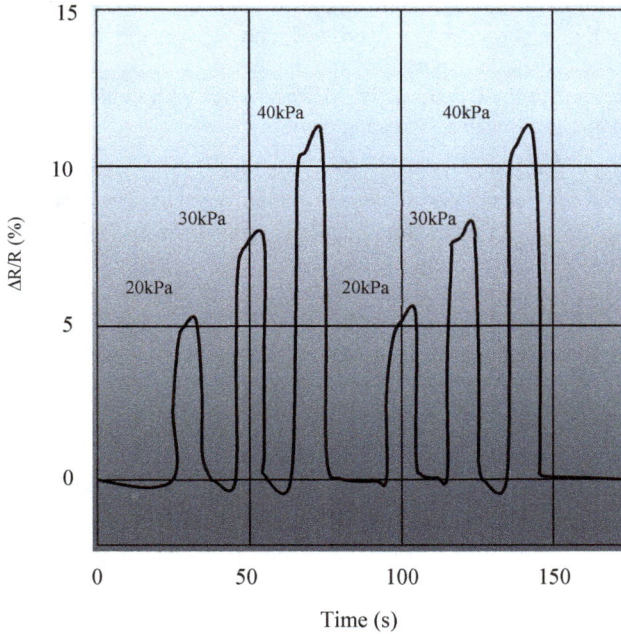

Figure 24. Reproducibility of the change in resistance during repeated cycles of pressure

It should be noted that current preparation methods can produce irregularly-shaped μm-sized droplets and lead to anisotropic properties, as in the stabilization of sub-μm sized droplets of EGaIn in polymer matrices. In the synthesis of EGaIn nanodroplets, stabilized by polymeric ligand encapsulation, surface-initiated atom-transfer radical polymerization initiation is used to functionalize the EGaIn nanodroplet oxide layer by means of polymethylmethacrylate and similar materials. The nanodroplets have good mechanical and thermal properties, and can exhibit a large undercooling; down to -80C. Such nanodroplets are stable in organic solvents, water or polymers at concentrations of up to 50wt% and can be direct solution-cast into flexible hybrid materials. The metal can be recovered from these dispersions by treatment with acid. Microcapsules which contain a liquid-metal core of gallium-indium can be prepared by urea-formaldehyde micro-encapsulation[256]. Ellipsoidal capsules, with major:minor diameter aspect-ratios ranging from 1.64 to 1.08 and having major diameters ranging from 245 to 3μm, have been produced. As the major diameter decreased, the aspect-ratio approached unity. The microcapsules survived incorporation into an epoxy matrix, and were triggered by mechanical damage to the cured matrix. Microcapsules with liquid-metal cores therefore have potential applications in self-healing devices.

One liquid-metal integrated system[257] has combined soft electronics with near-field communications for the purpose of human motion-sensing. The strain sensor, antenna and interconnections, were made from liquid metal and the overall device exhibited a gel-like stretchability. The patterning of the device, which produced a miniature (less than 2cm diameter) skin-attachable layout, was achieved by exploiting the selective wetting properties of the reduced Ga-21.5In-10%Sn alloy. The behaviours of the strain sensor and antenna under large uniaxial tensile, compressive and complex modes of deformation were determined (figures 22 to 24). The system was able to communicate with an external near-field reader that could transfers power to the device and also receive data wirelessly, and so without impairment even at a deformation of 30%.

As well as sensing human movement, similar methods can be used to reproduce it. An electrically responsive composite has been described[258] which exhibits muscle-like changes in elastic stiffness, of 1 to 10MPa, when stimulated using a voltage of 5 to 20V. The device consists of a stiffness-changing component which contains an embedded layer of conductive thermoplastic elastomer made up of propylene–ethylene copolymer and a percolating network of carbon-black. Opposite surfaces of the conductive thermoplastic elastomer layer are coated with a 20μm-thick film of EGaIn. When a voltage is applied to the EGaIn, an electric current passes through the conductive thermoplastic elastomer, causing Joule-heating and a consequent phase transition which changes the composite from its stiff state, with a Young's modulus of 10.4MPa, to its compliant state, having the much lower modulus of 0.7MPa. Differential scanning calorimetry shows that the change in state is governed by a solid–liquid transition. The voltage-dependent activation time can be less than 2s, and the composite can recover its original shape following large strains.

Conformable Electronics

The topics of reconfigurable antennae and sensors are closely related to the present topic, and often overlap. Elastic antennae make wearable electronics more comfortable and, like the reconfigurable antennae, are largely based upon elastomers and liquid metals. When stretched, the conducting parts develop crack patterns which do not affect their electrical conductivity or lead to fatigue. They can be simpler to manufacture than are the patch, coil and helical antennae. At the same time, built-in sensors can be very important in providing feedback. The connectors in wearable electronics applications should ideally have a high conductivity and exhibit a high degree of elastic deformability. A recently reported[259] liquid-metal filled elastic conductor combined an excellent conductivity (1.34 x 10^3S/cm) with an elongation-to-fracture of 116.86%; with the relative resistance variation being only 4.305% at that strain. The metal filling formed moreover a novel 3-dimensional conducting network, within the elastic matrix, which imparted excellent dynamic stability during stretching. One of the ultimate aims of this field is to interface biomedical electronic devices, android-style, directly with the human body in order to obtain precise medical data or to apply necessary electrical stimuli. This also extends logically to the use of liquid metals for implantable sensors. Hence the importance of creating flexible and stretchable electronic devices, especially those involving conducting polymers, liquid metal alloys and meandering architectures, which can adapt to the mechanics of soft tissues.

The microfluidic approach, based upon printed circuit board technology, thus permits the processing of large cross-section conductors and contacts which can support a large degree of stretching between embedded rigid active components and the remainder of the system. Tape transfer printing has been proposed as being a rapid prototyping process, in spite of a relatively low resolution of 150μm. Isolated patterns can be obtained by using a simple one-step process. As a demonstration[260], a radio-frequency identification tag was shown to be able to survive being stretched by 50% over 3000 times.

Deformable tactile sensing arrays that can simultaneously offer high spatial resolution, sensitivity and stretchability have been made[261] by using a casting process which produces arrays of microfluidic channels in low-modulus polymer membranes that can be as thin as 1mm. By then using liquid metal as a conductor, matrix-addressed capacitive sensing is used to resolve spatially-distributed strain, to mm-precision, over areas of several square centimetres. Because low-modulus polymers are used, these devices easily attain stretchabilities that are greater than 500%.

A pressurised liquid-metal screen-printing method has been proposed for the rapid manufacture of electronically conducting patterns[262]. Atomized liquid-metal microdroplets were pushed through a mesh so as to wet the substrate and produce the required liquid-metal pattern. The width and thickness of the printed tracks could be as small as 233.7 and 94.5μm, respectively, with the root-mean-square roughness of the printed surface being 1.27μm. A radio-frequency identification tag antenna could again be made in this manner. Microdroplets of gallium-based liquid metals have in fact found many applications in the fields of electronics and biomedicine but, due to the high surface tension of the liquid metal, high-quantity production of uniformly-sized liquid-metal microdroplets is difficult when using the usual acoustic or microfluidic methods. By modifying[263] the submerged electrodispersion technique which was normally used for generating water-based microdroplets, a simple method has been developed for the high-rate production of nearly monodisperse liquid-metal microdroplets in oil; without employing microfluidic techniques. The new method was capable of producing microdroplets with diameters ranging from tens to hundreds of μm, and a spinning disk was used to introduce a flow of oil and thus prevent coalescence of the microdroplets.

With their high conductivity and stretchability over large cross-sections, these liquid metals are ideal for creating stretchable devices and interconnects. On the other hand, creating high-resolution parallel features can be difficult and are usually limited to more than 100μm. Multi-level electroplated stencils have been investigated[264] for the printing of liquid metals and, in the case of galinstan, features as small as 10μm have been printed onto soft elastomers.

Highly stretchable electrical circuits were made[265] from EGaIn which was enclosed in elastomeric microfluidic channels. A particular microfluidic structure was made from polydimethylsiloxane and Ecoflex-0030 elastomers, which had differing stiffnesses. These circuits could be flexed, twisted and stretched to up to 2.2 times their original length. When such circuits were used for radio-frequency antennae, the latter suffered no degradation of reflected power even after being repeatedly stretched, to a tensile strain of 120%, more than 100 times[266]. The stretchability permitted the resonant frequency of the antenna to be mechanically tuned to about 1GHz.

A one-step liquid-metal transfer-printing method[267] which works well with a wide range of substrates involves just polymer-based adhesive, a printing machine, liquid-metal ink and a soft substrate. Even in the case of substrates which exhibit only a weak wettability with respect to liquid metals, this liquid-metal transfer-printing process still works well enough to create complex conducting geometries, multilayer circuits and large-area conducting patterns. The transfer-efficiency is excellent and the product exhibits a very good electrical stability, making it ideal for use in the field of wearable electronics.

Later stretchable conductors were developed[268] which consisted of liquid metal that had been injected into the microchannels of tri-block co-polymer gels, with rheological measurements being used to identify the temperature range within which the gel could be moulded and laminated, in order to form microchannels, without collapsing the microscale features. The best gel composition was a compromise between a minimal modulus, thus permitting maximum easy stretching, and maximum interfacial adhesive strength at the laminated polymer/polymer interface. The resultant 2-dimensional stretchable conductors could retain their electrical conductivity up to large strains with, in a typical case, conductivity persisting at up to at least 600% strain. Straining cycles could be repeated without any great degradation of the mechanical or electrical properties. The materials could also be easily recycled to yield full recovery of the components.

As mentioned in the Sensors section, a soft ultra-thin stretchable electronic skin, comprising surface-mounted components, was developed[269] which could be transferred to, and wrapped around, any 3-dimensional surface. It could also self-adhere to human skin. The 5μm-thick circuit was made by printing a pattern onto temporary tattoo paper by using a laser printer, and then coating it with silver ink and EGaIn. The resultant patterns were highly conducting and retained a low electrical resistivity as the circuit was stretched so as to conform to non-developable 3-dimensional surfaces. Surface-mounted micro-electronic chips could be integrated by means of a z-axis conducting interface which consisted of magnetically-aligned EGaIn-coated Ag-Ni microparticles, embedded in polyvinylalcohol. This conducting glue aided electrical contact with microchips having pins as small as 300μm. When printed onto temporary tattoo transfer paper, the circuit could be attached to a 3-dimensional surface via hydrographic transfer.

Clog-free oxide-free gallium-based liquid metal ink-jet printing has been available for some years for application to flexible electronics. Such printing can produce patterns in either conducting or non-conducting materials. In order to print the former, metal nano-particles may be dispersed in a solvent. A simple polydimethylsiloxane-based ink-jet printer, with hydrochloric acid-impregnated paper at the orifice, can produce a constant stream of gallium-based liquid metal droplets. Depending upon the flow-rate, pinch-off and Rayleigh instabilities may be observed[270]. Beads-on-string gallium-based liquid metal shapes could be produced on flexible substrates such as a silicon wafer, polydimethylsiloxane or paper. Ink-jet printed gallium-based liquid metal could maintain its line shape without disconnection, even in the presence of considerable deformation of flexible paper.

Most electronic skins are based upon organic or inorganic electronic materials, but inorganic materials are plagued by low flexibility while organic materials are plagued by limited conductivity. A new approach[271] is to make electronic tattoos from Ni-EGaIn material, which possesses a high (1.61 x 10^6S/m) conductivity as well as exhibiting human-skin compliance. The method is based upon the adhesion of semi-liquid Ni-EGaIn and polymethacrylate glue to human skin.

Problems persist in the preparation of stretchable electronics, one of which is to interface soft-circuit wiring with silicon chips. The other is the preparation of multi-layer circuits. One solution[272] is to provide vertical interconnection access by using liquid metal to fill cavities which are formed by laser ablation. This permits the production of reliable multi-layer stretchable circuits which can survive strains of over 80%.

Because gallium interacts metallurgically with most metals, use of these liquid alloys naturally leads to unstable interfaces when combined with metal electrodes or interconnects, thus impairing the easy integration of EGaIn into electronic circuitry. The present writer well remembers a mathematician colleague who wished to measure the meniscus interaction of small steel balls floating on liquid metal. Mercury being too toxic for this purpose, gallium-indium alloy was chosen instead. Without telling his metallurgist colleagues, he asked the workshop to make a sturdy container, in which to hold the several kg of liquid alloy, … it was made from aluminium. The result was all too predictable, and rather embarrassing, as the research group specialised in eutectic studies.

Perhaps due to such *ad hoc* eutectic formation, or to rapid grain-boundary penetration, gallium-based alloys can certainly have very corrosive effects upon various metals. One study[273] showed that galinstan starts to damage aluminium within 300s, and completely destroys small samples in about 0.5h, at room temperature. In the case of brass at room temperature there was no appreciable damage after 24h. At 100C, damage began after 300s. At 230C, a small sample was completely destroyed within 3h. Copper did not exhibit any damage after 24h of contact with galinstan at room temperature. At 230C, it required about 1h to dissolve completely. Not unexpectedly, stainless steel fared better, and washers were undamaged after 5h at 230C; apart from a change in colour. Again not unexpectedly, aluminium was not recommended for wiring purposes at any temperature.

In this context, printed graphene was demonstrated[274] to be a reliable high-performance interfacial layer for making electrical connections to EGaIn, without dissolving those connections. A circa 100nm graphene film, printed between silver leads and EGaIn, acted as a physical barrier which effectively passivated the surface with respect to alloying but retained the ability to conduct current across the interface. Graphene interfacial contacts were durable, and were thermally stable at up to 300C. They also resisted bending. In another study[275] of the integration of surface-mounted micro-electronic components with liquid-metal electrical interconnects, the latter were created by making copper patterns on a soft elastomer substrate and exposing that substrate to the EGaIn, which then selectively wetted the copper tracks. In order to ensure that a secure electromechanical connection existed between the EGaIn and the micro-electronics, the terminals of the metal-coated tracks were, in effect, soldered to the metal pins of the packaged micro-electronic circuits by using a hydrochloric-acid vapour-treatment. These techniques facilitated the creation of stretchable circuits which consisted of liquid-metal wiring and micro-electronic components. The hydrochloric-acid vapour-treatment markedly improved the electrical conductivity at the metal/pin interface as well as the strain-limit of the soft circuits.

Additive manufacturing techniques can be used to construct the mechanical and electronic parts of a system simultaneously[276]. This involves printing hollow channels, within 3-dimensional printed parts, which are then filled with a low melting-point liquid alloy that solidifies to form electrical tracks[277]. The method is compatible with many commercial stereolithography machines[278]. Additive manufacturing of metals and alloys generally permits the multi-material manufacturing of products, with high resolution and microstructural control. On the other hand, micro-additive manufacturing is more difficult than with plastics and ceramics due to the very different thermofluidic properties of metals. A recent study[279] of the flow and deposition of EGaIn focussed on the effect of the elastic solid skin of gallium oxides which encapsulates extruded filaments. Reduced adhesion between the nozzle and the oxide skin greatly improved process repeatability, and stress-analysis showed that the oxide skin suffered axial, circumferential and shear stresses during processing. Continuous filamentary extrusion was possible within a circumferential-stress range which was a function of the nozzle diameter and printing angle and was governed by the pressure used. During extrusion, the oxide skin continuously yielded axially while suffering a shear which depended upon the printing angle. The filament diameter was a function of the shear strain and the nozzle diameter. Decreasing the pressure during printing led to a reduction in the filament diameter, but increasing the pressure did not increase the filament diameter.

A recent novel approach[280] to the use of high-resolution reconfigurable 3-dimensional printing of liquid metals promises that a minimum line-width of 1.9mm can be reliably formed by means of direct printing, and that these patterns can be reconfigured into 3-dimensional structures having the same resolution. Such reconfiguration can be repeated, and moreover generates a thin oxide interface. The free-standing structures can then be encapsulated within a stretchable conforming and passivating material. The technique has applications in the fields of reconfigurable antennae and of interconnections which could be used as mechanical switches. A 12mm-wide EGaIn line was used to connect 2 pads of gold, copper or silver, with the length ranging from 5 to 150mm. The resistance increased linearly with length (figure 25). These 3-dimensional electrode arrangements could also minimize the required numbers and spacings of interconnections. The stability of electrodes with respect to electrical loading becomes important as device designs are miniaturized. In electrical breakdown tests, a pre-printed EGaIn line was reconfigured so as to connect 2 metal contact pads: 5nm-thick chromium and 500nm-thick gold. A direct-current or alternating-current (120Hz) bias was used to measure the resistance. The current density increased almost linearly with the applied electric field and then saturated at $1.9 \times 10^{10} A/m^2$ before breakdown. The alternating-current results exhibited a similar trend. This maximum current density of the EGaIn electrode was of the same order of magnitude (about $10^{10} A/m^2$) as that of vacuum-deposited solid metal films; but those are fragile and non-stretchable. As the direct-current electric field neared the breakdown condition of some 12V/mm, the EGaIn flowed to the negative pad and broke down due to electrohydrodynamic forces acting between the end of the positive pole, to the negative pole of the pads. Under alternating-current conditions near to the breakdown electric field, the EGaIn flowed toward both positive and negative poles and breakdown occurred at both ends. Gallium and indium atoms were widely dispersed around the electrically blown regions in both cases. When the mechanical stability of 3-dimensional EGaIn structures was tested at high temperatures, they could maintain their initial free-standing form during heating at 500C for 0.5h. Following repeated heating at 500C, and cooling to room temperature, only the oxide skin became slightly wrinkled, and this was attributed to differences in thermal expansion between the oxide and the EGaIn. The contact resistance between direct-printed EGaIn lines and gold, copper and silver were 68.2, 69.5 and 89.8ohm-mm, respectively. The equivalent values for reconfigured EGaIn were 70.2, 70.9 and 90.5ohm-mm, respectively. There was therefore a negligible increase in the resistance, regardless of the presence of a thin oxide skin. The oxide interface of reconfigured EGaIn electrodes retarded the penetration of gallium, from EGaIn, into metal layers and the possibility of liquid-metal embrittlement. As noted elsewhere, when EGaIn comes into direct contact with other metal layers without any oxide interface, gallium atoms can penetrate their grain boundaries, change the interfacial energy and accelerate intergranular failure.

A further intriguing possibility is that of producing *transparent* stretchable metal-grid electrodes. By using a so-called roll-painting and lift-off technique, based upon photolithography, stretchable EGaIn grids having a linewidth of 20μm and a line-separation of 400 to 1000μm could be produced[281]. These had a transmittance of 75 to 88%, and a sheet-resistance of less than 2.3Ω/square. They also provided stable conductivity during stretching by up to 40%, and a high reliability under cyclic deformation. By combining these EGaIn grids with transparent organic coatings, it was possible to create highly deformable and uniformly conducting transparent electrodes for use as a stretchable inorganic electroluminescence device.

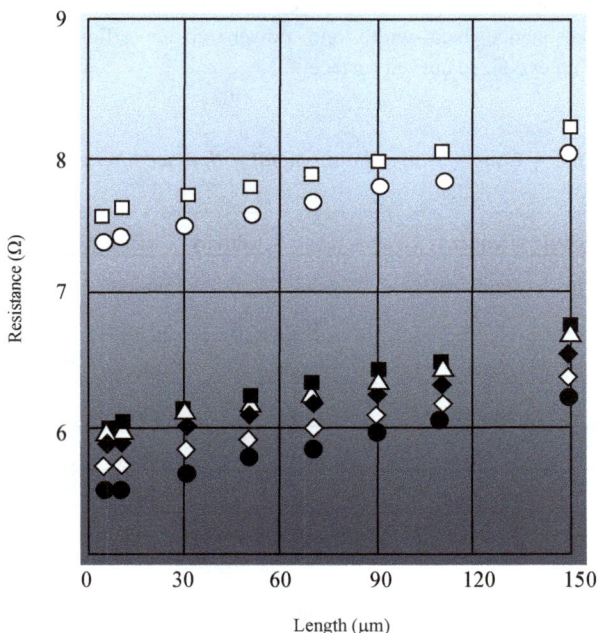

Figure 25. Dependence of the total resistance of EGaIn lines upon the length of the channel. From top to bottom, the data are for reconfigured lines between silver, direct-printed lines between silver, reconfigured between copper, direct-printed between copper, reconfigured between gold and direct-printed between gold

Telemetry devices used for implanted medical sensors are considerably affected by their environment, given that living tissue is a frequency-dependent lossy dielectric. Soft flexible coils made from liquid metals encased in biocompatible elastomers can be envisaged for use in telemetry systems that are part of implantable medical devices. A coil offering high conductivity, minimal losses and a high unloaded Q-factor, is required for use in an efficient wireless telemetry system. The conductivity of EGaIn is about one tenth of that of gold or copper, but it is nevertheless possible to create efficient biomedical telemetry systems by using liquid-metal coils. A wireless telemetry system for an artificial retina has been used[282] to test such liquid-metal coils. Power-transfer efficiencies of 43% and 21% were obtained at inter-coil operating distances of 5 and 12mm, respectively. The liquid-metal based coil still retained more than 72% of its voltage-gain, resonance band-width and power-transfer efficiency when physically deformed over an eye-sized curved surface.

Table 4. Comparison of liquid-metal with other tuning techniques

Method	Speed	Frequency	Temperature Sensitivity	Bias	Power	Cost
Present	ms	unlimited	high	unneeded	50mW	low
YIG	ms	limited	high	magnetic	0.5-5W	high
BST	ns	limited	high	electric	0	high
Diode[S]	ns	limited	low	electric	0	low
Diode[P]	ns	limited	low	electric	20-30mA	low
MEMS	µs	limited	low	electric	0	high

S: Schottky-type, P: pin-type

Even more intrusive applications are found in the field of so-called biobotics, where neural or muscular stimulation is introduced via implanted electrodes in order to control animal movement directly. A degradation of the efficiency of stimulation can occur after extended use, and this is partly due to an incompatibility of the implanted electrodes. Plain stainless-steel wires were originally used as electrodes, but it was proposed here[283] that EGaIn could be used to enhance the properties of stainless-steel electrodes for *in vivo* neurostimulation.

An early proposed[284] fluidic microstrip band-stop filter exhibited transmission properties that changed via discrete states. It comprised EGaIn contained in microfluidic channels, with the former's fluidity permitting the open stub resonator of the filter to change its length by flowing under an applied pressure. A series of posts in the channel defined the length of stub which was filled by the metal, and determined the pressure which was required to be applied in order to cause the liquid metal to flow and thus increase the stub

length. The frequency-response of the filter then changed due to the changes in length of
the resonator stub.

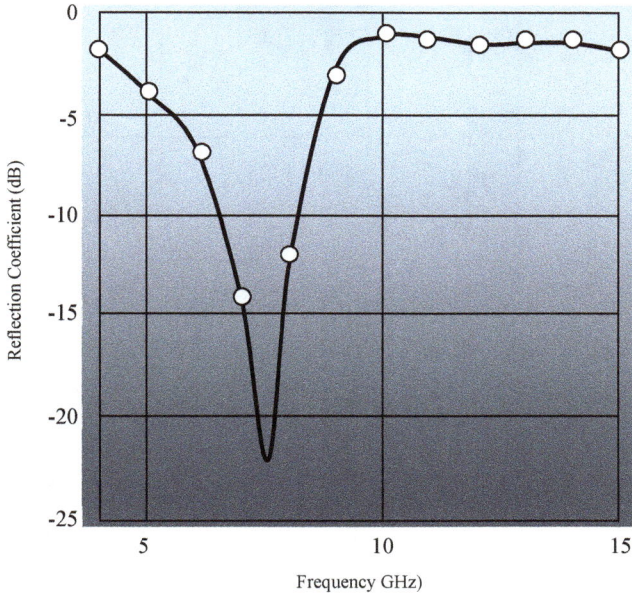

*Figure 26. Absorptivity of a switchable metamaterial absorber with injected liquid metal.
The absorptivity is hardly affected as the resistance ranges from 99 to 101Ω.*

A switchable band-pass/band-stop filter design has recently been proposed[285] which uses
liquid metal as a fluidic switch. The design was based upon the Chebyshev response, and
involved using a 3-stage quarter-wavelength resonant structure. The fluidic switch was
made by injecting EGaIn into microfluidic stubs which were encased in
polydimethylsiloxane. When the fluidic switch selected a short stub by using a
micropump and microprocessor, the filter acted as a band-pass filter with short stubs.
When the fluidic switch selected the open stub, the filter acted as a band-stop filter with
open stubs. In band-pass filter mode, the center frequency was 2.5GHz and the 1dB
bandwidth was 1.75 to 3.07GHz, with an insertion-loss of 0.5dB. In band-pass filter
mode, the 15dB bandstop bandwidth was 2.4 to 2.65GHz, with a 2.5GHz center

frequency. The use of liquid metals in this way was expected to confer several advantages when compared with other tuning techniques (table 4).

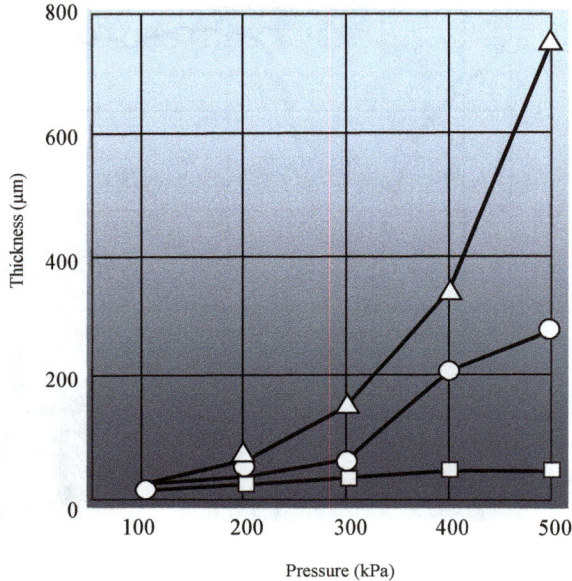

Figure 27. Liquid-metal thickness, deposited from 5cm, as a function of spraying pressure Triangles: 5s, circles: 3s, squares: 1s

Metamaterials are artificial structures which consist of periodically arranged metallic features and can exhibit counterintuitive properties such as a negative refractive index; thus making them of interest with regard to microwave applications. For example, a metamaterial absorber can be very thin but, on the other hand, it may have only a narrow bandwidth because of its electric and magnetic resonances. A frequency-tunable metamaterial absorber can compensate for bandwidth limitations, and can receive electromagnetic waves from any direction. Most tunable metamaterial absorbers are made by incorporating electronic tuning components, such as varactor diodes, and the relative bandwidth of the metamaterial absorber can be extended by 30% because of frequency-tunability. The absorption ratio can be changed by electrostatically influencing the metamaterial layers. It was demonstrated[286] that a class of frequency-switchable metamaterial absorbers in the X-band could be created by injecting EGaIn into a microfluidic channel engraved on polymethylmethacrylate so as to permit frequency-switching by filling or emptying the microfluidic channels.

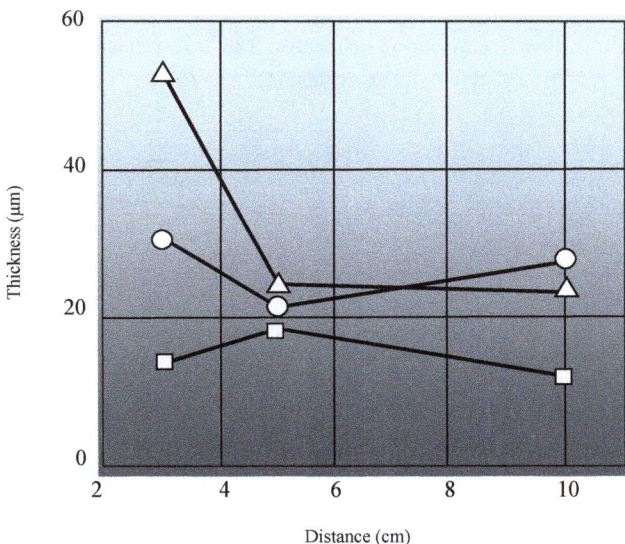

Figure 28. Liquid-metal line-thickness as a function of spraying-distance (100kPa) Triangles: 5s, circles: 3s, squares: 1s

When combined with a rectangular waveguide, the resonant frequency could be switched from 10.96 to 10.61GHz by injecting the liquid metal, while maintaining an absorption level of better than 98%. Metamaterial absorbers can reduce radar cross-sections, and tend to be thin, light and relatively cheap. Resonant metamaterial absorbers unfortunately suffer from the drawback of a narrow bandwidth when, for radar applications, wide-band absorption is required. A wide-band switchable metamaterial absorber was proposed[287,288] which used liquid metal but, in order to be able to reduce the radar cross-section in both the X-band and C-band, a switchable so-called Jerusalem-cross resonator was designed. The resonator consisted of slotted circular rings, chip resistors and microfluidic channels. It was etched onto a flexible printed-circuit board and microfluidic channels were laser-etched into polydimethylsiloxane. This absorber could have its absorption frequency band switched by injecting liquid metal into the channels. Its behaviour was demonstrated using full-wave simulations and experimentation. The latter results indicated absorption ratios of over 90%, between 7.43 and 14.34GHz and between 5.62 to 7.3GHz, for empty channels and metal-filled channels, respectively. The absorption band was thus successfully switched between the C-band (4 to 8GHz) and X-band (8 to 12GHz) by injecting EGaIn into the channels. The absorptivity when the channels were metal-filled was not affected by the resistance (figure 26).

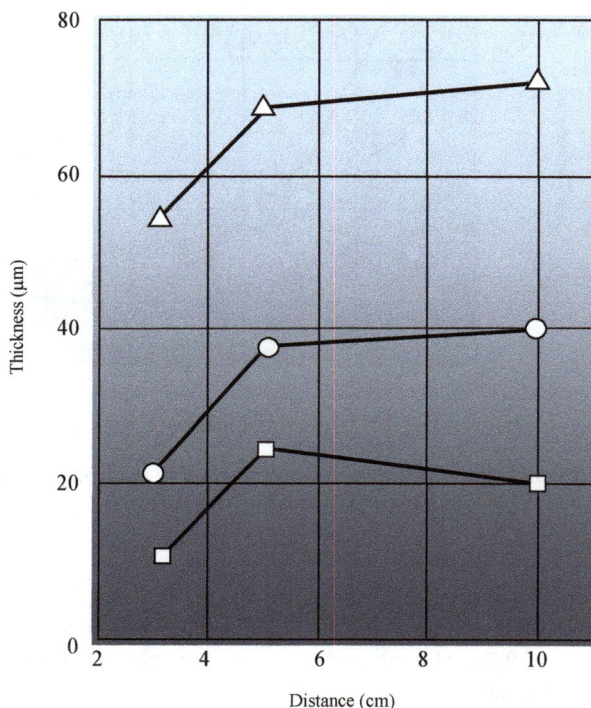

*Figure 29. Liquid-metal line-thickness as a function of spraying-distance (200kPa)
Triangles: 5s, circles: 3s, squares: 1s*

An EGaIn liquid-metal electrode was designed[289] to be used with polydimethylsiloxane in order to create a dielectric elastomer actuator for soft-matter applications. This solved the previous persistent problem of electrode choice and deposition. The liquid metal offered both a high conductivity and complete soft-matter compatibility in that the liquid electrode had almost the same electrical conductivity as solid-metal wiring but did not mechanically impede bending or stretching of the dielectric elastomer actuator. This innovation led to the successful actuation of sample curved cantilever elastomer actuators. Like most electrostatically-controlled elastomer-based devices, the voltage requirements of the curved dielectric elastomer actuators were however considerable.

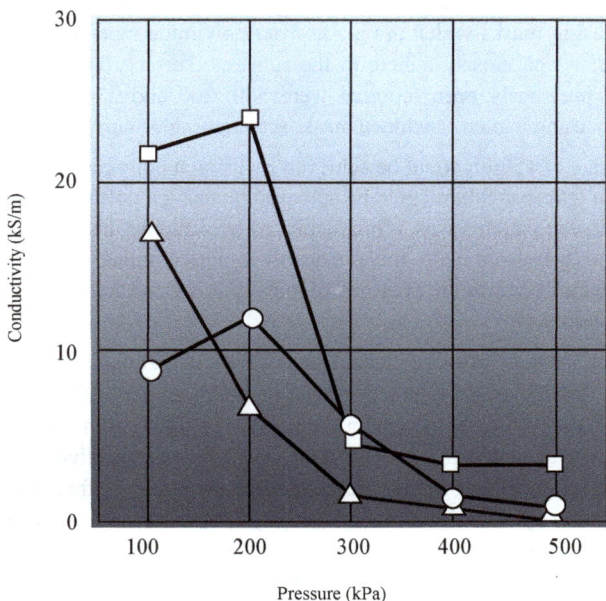

Figure 30. Liquid-metal conductivity as a function of spray pressure at 10cm Triangles: 5s, circles: 3s, squares: 1s

Spray-coated liquid metals are compatible with elastic substrates, and ultraviolet-curable and polyimide masks can be used to pattern the sprayed metal. The average droplet diameters were 2.435µm at 100kPa, 1.507µm at 300kPa and 0.715µm at 500kPa, with standard deviations of 0.2546, 0.1773 and 0.1295µm, respectively. The effect of the spraying parameters upon the thickness (figures 27 to 30) and conductivity of the liquid metal has been characterized in great detail[290]. A minimum linewidth of 48µm and a minimum gap-width of 34μm can be achieved. A liquid-metal patch antenna and transmission line have been constructed by using this method. But although the spraying of liquid metal requires only small amounts of metal for the creation of conductive patterns, the preparation of micromachined shadow masks with which to pattern the liquid increases the fabrication time. This problem has been circumvented by using polyimide tape and polyurethane acrylate masks; thus reducing the preparation time to less than 300s. Another problem which was associated with the spraying of liquid metal was the ingress of liquid metal below the edges of stencil masks. This difficulty has been solved by using masks which are made from polyimide tape and polyurethane acrylate membranes, which closely adhere to the substrate surface. The minimum feature sizes which had previously been reported were 500, 200 and 234μm, and those had been achieved by using a micromachined mask, screen mesh or vinyl mask, respectively.

A feature-size of 150μm could be achieved by using a polyurethane acrylate mask, and a linewidth of less than 50μm could be achieved by using a polyimide mask. The minimum spacing between patterns was 34μm, achieved by scraping the sprayed liquid metal with a needle, as compared with the previously reported minimum of 75μm. This is an important factor because the presence of gaps of this size can be important when dealing with THz circuits.

Both of the present masking techniques were compatible with flexible and curved substrates. The thickness of the oxide layer on galinstan was estimated to be 1 to 3nm so, taking the higher estimate, the volume fractions of oxide in films sprayed at 100, 300 and 500kPa were calculated to be 0.37, 0.60 and 1.25%, respectively. Liquid-metal films which were sprayed using higher pressures contained a larger fraction of oxidized metal; thus resulting in a lower electrical conductivity. The conductivity consequently decreased as the thickness of the film increased; a result which is rather counter-intuitive. The air pressure was sufficiently strong however to break the oxide covering of the liquid which had already been deposited, so that newly-exposed liquid was rapidly oxidized. The amount of oxide in the film thus increased and decreased the conductivity of thicker films, with their longer spraying times. The liquid metal was sprayed onto a nitrile surface in order to determine the effect of deformation upon the resistance of stretched films. As the strain increased, the resistance of the film increased (figure 31) … up to a strain of 80%. When it was stretched beyond 80%, the resistance suddenly increased by 3 orders of magnitude. On the other hand, the same effect had been reported for metal films and for Ag–Au nanowire electrodes on an elastomer substrate, and the sudden increase there had occurred after strains of only 10 and 30%, respectively.

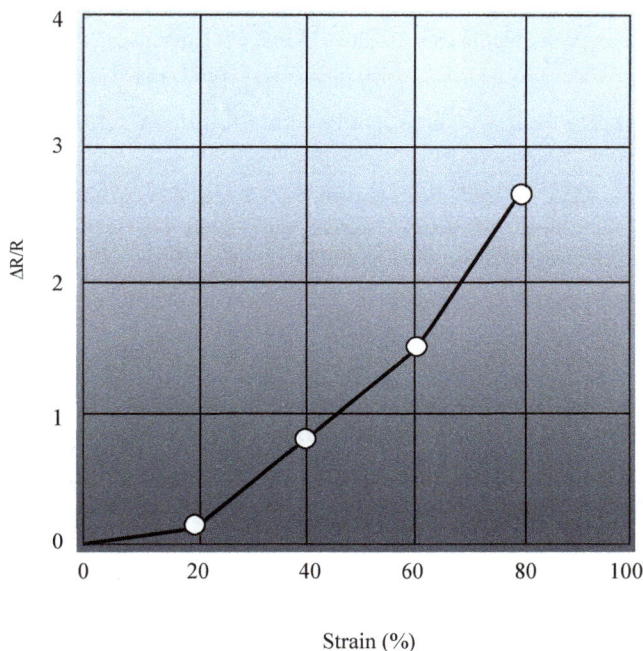

Figure 31. Change in resistance of a sprayed liquid-metal pattern with uniaxial strain

It is logical that stretchable and conformable electronic devices may well require batteries which share their capabilities. A rechargeable stretchable battery has been designed[291] which comprises a EGaIn anode, carbon paste, a MnO_2 slurry cathode, an alkaline electrolytic hydrogel and soft elastomeric packaging. It can cycle between 1.40 and 1.86V at $1mA/cm^2$ even while being subjected to a tensile strain of 100%, and this is possible due to a mechanism which involves the reversible stripping and plating of gallium, plus MnO_2 chemical conversion. The contact-area between the EGaIn anode and the hydrogel is increased by using $CaCl_2$ additives which reduce polarization and thereby reduce the effective current density. This then leads to higher discharge-plateaux and to lower charge-plateaux. This battery has an areal specific capacity of about $3.8mAh/cm^2$, and supports more than 100 charging-cycles. It can also be bent to a 2mm radius-of-curvature, with only a negligible change occurring in the electrochemical properties.

References

[1]   German patent application 102009013565, 5th November 2009.

[2]   Anon., The Manufacturer and Builder, 11[1], 1879, 134

[3]   Jia, M., Newberg, J.T., Journal of Physical Chemistry C, 123[47] 2019, 28688-28694. https://doi.org/10.1021/acs.jpcc.9b07731

[4]   Dickey, M.D., Chiechi, R.C., Larsen, R.J., Weiss, E.A., Weitz, D.A., Whitesides, G.M., Advanced Functional Materials, 18[7] 2008, 1097-1104. https://doi.org/10.1002/adfm.200701216

[5]   Kramer, R.K., Boley, J.W., Stone, H.A., Weaver, J.C., Wood, R.J., Langmuir, 30[2] 2014, 533-539. https://doi.org/10.1021/la404356r

[6]   Yalcintas, E.P., Ozutemiz, K.B., Cetinkaya, T., Dalloro, L., Majidi, C., Ozdoganlar, O.B., Advanced Functional Materials, 1906551, 2019. https://doi.org/10.1002/adfm.201906551

[7]   Zhu, B., Cai, Y., Huang, H., Liang, X., Li, X., Yang, H., IEEE/ASME International Conference on Advanced Intelligent Mechatronics, AIM, 2019-July, 8868474, 2019, 1652-1657.

[8]   Park, M., Im, J., Shin, M., Min, Y., Park, J., Cho, H., Park, S., Shim, M.B., Jeon, S., Chung, D.Y., Bae, J., Park, J., Jeong, U., Kim, K., Nature Nanotechnology, 7[12] 2012, 803-809. https://doi.org/10.1038/nnano.2012.206

[9]   Koh, A., Hwang, W., Y. Zavalij, P., Chun, S., Slipher, G., Mrozek, R., Materialia, 8, 2019, 100512. https://doi.org/10.1016/j.mtla.2019.100512

[10] Zhu, J.Y., Tang, S.Y., Khoshmanesh, K., Ghorbani, K., ACS Applied Materials and Interfaces, 8[3] 2016, 2173-2180. https://doi.org/10.1021/acsami.5b10769

[11] Yang, T., Foulkes, T., Kwon, B., Kang, J.G., Braun, P.V., King, W.P., Miljkovic, N., IEEE Transactions on Components, Packaging and Manufacturing Technology, 9[12] 2019, 2341-2351. https://doi.org/10.1109/TCPMT.2019.2930089

[12] Prabu, T., Sekar, C.A., Journal of Solar Energy Engineering, Transactions of the ASME, 139[6] 2017, 064501. https://doi.org/10.1115/1.4037821

[13] Zhang, X.D., Yang, X.H., Zhou, Y.X., Rao, W., Gao, J.Y., Ding, Y.J., Shu, Q.Q., Liu, J., Energy Conversion and Management, 185, 2019, 248-258. https://doi.org/10.1016/j.enconman.2019.02.010

[14] Zhang, R., Hodes, M., Lower, N., Wilcoxon, R., IEEE Transactions on

Components, Packaging and Manufacturing Technology, 5[6] 2015, 762-770. https://doi.org/10.1109/TCPMT.2015.2426791

[15]   Lam, L.S., Hodes, M., Enright, R., Journal of Heat Transfer, 137[9] 2015, 091003. https://doi.org/10.1115/1.4030208

[16]   Muhammad, A., Selvakumar, D., Wu, J., International Journal of Heat and Mass Transfer, 150, 2020, 119265. https://doi.org/10.1016/j.ijheatmasstransfer.2019.119265

[17]   Li, Z., Li, J., Li, X., Ni, M.J., Experimental Thermal and Fluid Science, 82, 2017, 240-248. https://doi.org/10.1016/j.expthermflusci.2016.11.021

[18]   Kim, D., Lee, D.W., Choi, W., Lee, J.B., Proceedings of the IEEE International Conference on Micro Electro Mechanical Systems, 6170184, 2012, 1005-1008.

[19]   Li, G., Parmar, M., Lee, D.W., Proceedings of the IEEE Conference on Nanotechnology, 6721059, 2013, 410-413.

[20]   Ilyas, N., Butcher, D.P., Durstock, M.F., Tabor, C.E., Advanced Materials Interfaces, 3[9] 2016, 1500665. https://doi.org/10.1002/admi.201500665

[21]   Kim, D., Lee, D.W., Choi, W., Lee, J.B., Journal of Microelectromechanical Systems, 22[6] 2013, 1267-1275. https://doi.org/10.1109/JMEMS.2013.2278625

[22]   Kim, D., Thissen, P., Viner, G., Lee, D.W., Choi, W., Chabal, Y.J., Lee, J.B., ACS Applied Materials and Interfaces, 5[1] 2013, 179-185. https://doi.org/10.1021/am302357t

[23]   Shang, C., Yang, J., Zhang, J., Ni, M., Chinese Journal of Theoretical and Applied Mechanics, 51[2] 2019, 380-391.

[24]   Davis, E., Ndao, S., Advanced Engineering Materials, 20[3] 2018, 1700829. https://doi.org/10.1002/adem.201700829

[25]   Koh, A., Mrozek, R., Slipher, G., Advanced Materials Interfaces, 5[5] 2018, 1701240. https://doi.org/10.1002/admi.201701240

[26]   Liu, T., Sen, P., Kim, C.J., Journal of Microelectromechanical Systems, 21[2] 2012, 443-450. https://doi.org/10.1109/JMEMS.2011.2174421

[27]   Liu, T., Sen, P., Kim, C.J., Proceedings of the IEEE International Conference on Micro Electro Mechanical Systems, 5442440, 2010, 560-563.

[28]   Kadlaskar, S.S., Yoo, J.H., Abhijeet, Lee, J.B., Choi, W., Journal of Colloid and Interface Science, 492, 2017, 33-40. https://doi.org/10.1016/j.jcis.2016.12.061

[29] Knoblauch, M., Hibberd, J.M., Gray, J.C., Van Bel, A.J.E., Nature Biotechnology, 17[9] 1999, 906-909. https://doi.org/10.1038/12902

[30] Sutter, P.W., Sutter, E.A., Nature Materials, 6[5] 2007, 363-366. https://doi.org/10.1038/nmat1894

[31] Guan, Y., Wang, S., Tentzeris, M.M., Liu, Y., IEEE MTT-S International Microwave Symposium Digest, 8701082, 2019, 1446-1449.

[32] Channaa, H., Surmann, P., Pharmazie, 64[3] 2009, 161-165.

[33] Surmann, P., Channaa, H., Electroanalysis, 27[7] 2015, 1726-1732. https://doi.org/10.1002/elan.201400752

[34] Jedlińska, K., Wegiel, K., Baś, B., Journal of the Electrochemical Society, 165[14] 2018, B708-B712. https://doi.org/10.1149/2.0691814jes

[35] Mellor, B.L., Kellis, N.A., Mazzeo, B.A., Review of Scientific Instruments, 82[4] 2011, 046110. https://doi.org/10.1063/1.3581229

[36] Huang, R., Xiong, X., Zhu, E., Journal of Mechanical Engineering, 55[9] 2019, 176-182. https://doi.org/10.3901/JME.2019.09.176

[37] Schreiber, S., Minute, M., Tornese, G., Giorgi, R., Duranti, M., Ronfani, L., Barbi, E., Pediatric Emergency Care, 29[2] 2013, 197-199. https://doi.org/10.1097/PEC.0b013e3182809c29

[38] Ahlberg, P., Jeong, S.H., Jiao, M., Wu, Z., Jansson, U., Zhang, S.L., Zhang, Z.B., IEEE Transactions on Electron Devices, 61[8] 2014, 2996-3000. https://doi.org/10.1109/TED.2014.2331893

[39] Hoshyargar, F., Khan, H., Kalantar-Zadeh, K., O'Mullane, A.P., Chemical Communications, 51[74] 2015, 14026-14029. https://doi.org/10.1039/C5CC05246G

[40] Oloye, O., Tang, C., Du, A., Will, G., O'Mullane, A.P., Nanoscale, 11[19] 2019, 9705-9715. https://doi.org/10.1039/C9NR02458A

[41] Hoshyargar, F., Crawford, J., O'Mullane, A.P., Journal of the American Chemical Society, 139[4] 2017, 1464-1471. https://doi.org/10.1021/jacs.6b05957

[42] Zoellner, B., Hou, F., Carbone, A., Kiether, W., Markham, K., Cuomo, J., Maggard, P.A., ACS Omega, 3[12] 2018, 16409-16415. https://doi.org/10.1021/acsomega.8b02442

[43] Bai, P., Li, S., Tao, D., Jia, W., Meng, Y., Tian, Y., Tribology International, 128, 2018, 181-189. https://doi.org/10.1016/j.triboint.2018.07.036

[44]   Le Brun, N., Markides, C.N., International Journal of Thermophysics, 36[10-11] 2015, 3222-3238. https://doi.org/10.1007/s10765-015-1986-0

[45]   Oh, J.H., Woo, J.Y., Jo, S., Han, C.S., Sensors and Actuators, A, 299, 2019, 111610. https://doi.org/10.1016/j.sna.2019.111610

[46]   Dickey, M.D., Chiechi, R.C., Larsen, R.J., Weiss, E.A., Weitz, D.A., Whitesides, G.M., Advanced Functional Materials, loc. cit.

[47]   Chen, Z., Lee, J.B., Proceedings of the IEEE International Conference on Micro Electro Mechanical Systems, 8870886, 2019, 396-399.

[48]   Lin, Y., Liu, Y., Genzer, J., Dickey, M.D., Chemical Science, 8[5] 2017, 3832-3837. https://doi.org/10.1039/C7SC00057J

[49]   Reineck, P., Lin, Y., Gibson, B.C., Dickey, M.D., Greentree, A.D., Maksymov, I.S., Scientific Reports, 9[1] 2019, 5345. https://doi.org/10.1038/s41598-019-41789-8

[50]   Bhardwaj, A., Verma, S.S., Optics Communications, 452, 2019, 264-272. https://doi.org/10.1016/j.optcom.2019.07.049

[51]   Griffin, A.S., Sottos, N.R., White, S.R., Proceedings of SPIE, 2018, 1059611.

[52]   Liu, N., Jin, Y., Miao, M., Cui, X., 17th International Conference on Electronic Packaging Technology, 7583325, 2016, 1135-1139.

[53]   He, Z.Z., Xue, X., Liu, J., ASME International Mechanical Engineering Congress and Exposition, Proceedings, 2014, 8B.

[54]   Luo, M., Liu, J., Frontiers in Energy, 7[4] 2013, 479-486. https://doi.org/10.1007/s11708-013-0277-3

[55]   De Castro, I.A., Chrimes, A.F., Zavabeti, A., Berean, K.J., Carey, B.J., Zhuang, J., Du, Y., Dou, S.X., Suzuki, K., Shanks, R.A., Nixon-Luke, R., Bryant, G., Khoshmanesh, K., Kalantar-Zadeh, K., Daeneke, T., Nano Letters, 17[12] 2017, 7831-7838. https://doi.org/10.1021/acs.nanolett.7b04050

[56]   Li, F., Kuang, S., Li, X., Shu, J., Li, W., Tang, S.Y., Zhang, S., Advanced Materials Technologies, 4[3] 2019, 1800694. https://doi.org/10.1002/admt.201800694

[57]   Chen, X., Chen, Q., Wu, D., Zheng, Y., Zhou, Z., Zhang, K., Lv, W., Zhao, Y., Lin, L., Sun, D., Journal of Raman Spectroscopy, 49[8] 2018, 1301-1310. https://doi.org/10.1002/jrs.5393

[58]   Musgrave, C.S.A., Lu, N., Sato, R., Nagai, K., RSC Advances, 9[24] 2019, 13927-

13932. https://doi.org/10.1039/C9RA01905G

[59] Tostmann, H., Dimasi, E., Shpyrko, O.G., Pershan, P.S., Ocko, B.M., Deutsch, M., Physical Chemistry Chemical Physics, 102[9] 1998, 1136-1141. https://doi.org/10.1002/bbpc.19981020913

[60] Dobosz, A., Daeneke, T., Zavabeti, A., Zhang, B.Y., Orrell-Trigg, R., Kalantar-Zadeh, K., Wójcik, A., Maziarz, W., Gancarz, T., Nanomaterials, 9[2] 2019, 235. https://doi.org/10.3390/nano9020235

[61] Kim, D., Yoo, J.H., Lee, J.B., Journal of Micromechanics and Microengineering, 26[4] 2016, 045004. https://doi.org/10.1088/0960-1317/26/4/045004

[62] Kim, D., Hwang, J., Choi, Y., Kwon, Y., Jang, J., Yoon, S., Choi, J., Cancers, 11[11] 2019, 1666. https://doi.org/10.3390/cancers11111666

[63] Morris, N.J., Farrell, Z.J., Tabor, C.E., Nanoscale, 11[37] 2019, 17308-17318. https://doi.org/10.1039/C9NR06369B

[64] Farzbod, F., Naghdi, M., Goggans, P.M., Journal of Vibration and Acoustics, Transactions of the ASME, 141[1] 2019, 014501. https://doi.org/10.1115/1.4041140

[65] Wang, Y., Duan, W., Zhou, C., Liu, Q., Gu, J., Ye, H., Li, M., Wang, W., Ma, X., Advanced Materials, 1905067, 2019. https://doi.org/10.1002/adma.201905067

[66] Pekas, N., Zhang, Q., Juncker, D., Journal of Micromechanics and Microengineering, 22[9] 2012, 097001. https://doi.org/10.1088/0960-1317/22/9/097001

[67] Yan, H., Ji, Y., Yan, J., Journal of Materials Science - Materials in Electronics, 30[16] 2019, 15766-15771. https://doi.org/10.1007/s10854-019-01962-1

[68] Yan, H., Ji, Y., Wang, Z., Yangbo, D., Wenbin, C., Guoyou, W., Tian, B., Ma, H., International Heat Transfer Conference, 2018, 5349-5356.

[69] Li, G., Ji, Y., Sun, Y., Ma, H., Xing, F., Liu, Y., Journal of Xian Jiaotong University, 50[9] 2016, 61-65.

[70] Kurz, W., Fisher, D.J., Trivedi, R., International Materials Reviews, 64[6] 2019, 311-354. https://doi.org/10.1080/09506608.2018.1537090

[71] Kurz, W., Fisher, D.J., Fundamentals of Solidification, Trans Tech Publications, Aedermannsdorf, Switzerland, 1989.

[72] Schurmann, D., Willers, B., Eckert, S., IOP Conference Series - Materials Science and Engineering, 424[1] 2018, 012005. https://doi.org/10.1088/1757-

899X/424/1/012005

[73] Mäder, K., Nauber, R., Beyer, H., Klass, A., Thieme, N., Büttner, L., Czarske, J., Technisches Messen, 82[11] 2015, 578-584. https://doi.org/10.1515/teme-2015-0084

[74] Dubovikova, N., Resagk, C., Karcher, C., Kolesnikov, Y., Measurement Science and Technology, 27[5] 2016, 055102. https://doi.org/10.1088/0957-0233/27/5/055102

[75] Zürner, T., Ratajczak, M., Wondrak, T., Eckert, S., Measurement Science and Technology, 28[11] 2017, 115301. https://doi.org/10.1088/1361-6501/aa7f58

[76] Zürner, T., Vogt, T., Resagk, C., Eckert, S., Schumacher, J., Experiments in Fluids, 59[1] 2018, 3. https://doi.org/10.1007/s00348-017-2457-0

[77] Forbriger, J., Galindo, V., Gerbeth, G., Stefani, F., Measurement Science and Technology, 19[4] 2008, 045704. https://doi.org/10.1088/0957-0233/19/4/045704

[78] Stefani, F., Gundrum, T., Gerbeth, G., Rüdiger, G., Szklarski, J., Hollerbach, R., New Journal of Physics, 9, 2007, 295. https://doi.org/10.1088/1367-2630/9/8/295

[79] Eaker, C.B., Hight, D.C., O'Regan, J.D., Dickey, M.D., Daniels, K.E., Physical Review Letters, 119[17] 2017, 174502. https://doi.org/10.1103/PhysRevLett.119.174502

[80] Fracasso, D., Valkenier, H., Hummelen, J.C., Solomon, G.C., Chiechi, R.C., Journal of the American Chemical Society, 133[24] 2011, 9556-9563. https://doi.org/10.1021/ja202471m

[81] Cademartiri, L., Thuo, M.M., Nijhuis, C.A., Reus, W.F., Tricard, S., Barber, J.R., Sodhi, R.N.S., Brodersen, P., Kim, C., Chiechi, R.C., Whitesides, G.M., Journal of Physical Chemistry C, 116[20] 2012, 10848-10860. https://doi.org/10.1021/jp212501s

[82] Thuo, M.M., Reus, W.F., Simeone, F.C., Kim, C., Schulz, M.D., Yoon, H.J., Whitesides, G.M., Journal of the American Chemical Society, 134[26] 2012, 10876-10884. https://doi.org/10.1021/ja301778s

[83] Simeone, F.C., Yoon, H.J., Thuo, M.M., Barber, J.R., Smith, B., Whitesides, G.M., Journal of the American Chemical Society, 135[48] 2013, 18131-18144. https://doi.org/10.1021/ja408652h

[84] Nijhuis, C.A., Reus, W.F., Barber, J.R., Whitesides, G.M., Journal of Physical Chemistry C, 116[26] 2012, 14139-14150. https://doi.org/10.1021/jp303072a

[85] Carlotti, M., Degen, M., Zhang, Y., Chiechi, R.C., Journal of Physical Chemistry C, 120[36] 2016, 20437-20445. https://doi.org/10.1021/acs.jpcc.6b07089

[86] Liao, K.C., Yoon, H.J., Bowers, C.M., Simeone, F.C., Whitesides, G.M., Angewandte Chemie, 53[15] 2014, 3889-3893. https://doi.org/10.1002/anie.201308472

[87] Yoon, H.J., Bowers, C.M., Baghbanzadeh, M., Whitesides, G.M., Journal of the American Chemical Society, 136[1] 2014, 16-19. https://doi.org/10.1021/ja409771u

[88] Bowers, C.M., Liao, K.C., Yoon, H.J., Rappoport, D., Baghbanzadeh, M., Simeone, F.C., Whitesides, G.M., Nano Letters, 14[6] 2014, 3521-3526. https://doi.org/10.1021/nl501126e

[89] Jiang, L., Sangeeth, C.S.S., Nijhuis, C.A., Journal of the American Chemical Society, 137[33] 2015, 10659-10667. https://doi.org/10.1021/jacs.5b05761

[90] Chen, J., Giroux, T.J., Nguyen, Y., Kadoma, A.A., Chang, B.S., Vanveller, B., Thuo, M.M., Physical Chemistry Chemical Physics, 20[7] 2018, 4864-4878. https://doi.org/10.1039/C7CP07531F

[91] Wimbush, K.S., Fratila, R.M., Wang, D., Qi, D., Liang, C., Yuan, L., Yakovlev, N., Loh, K.P., Reinhoudt, D.N., Velders, A.H., Nijhuis, C.A., Nanoscale, 6[19] 2014, 11246-11258. https://doi.org/10.1039/C4NR02933J

[92] Kumar, S., Van Herpt, J.T., Gengler, R.Y.N., Feringa, B.L., Rudolf, P., Chiechi, R.C., Journal of the American Chemical Society, 138[38] 2016, 12519-12526. https://doi.org/10.1021/jacs.6b06806

[93] Fisher, D.J., Materials Science Foundations, 52, 2019, 1-152.

[94] Barbee, M.H., Mondal, K., Deng, J.Z., Bharambe, V., Neumann, T.V., Adams, J.J., Boechler, N., Dickey, M.D., Craig, S.L., ACS Applied Materials and Interfaces, 10[35] 2018, 29918-29924. https://doi.org/10.1021/acsami.8b09130

[95] Barber, J.R., Yoon, H.J., Bowers, C.M., Thuo, M.M., Breiten, B., Gooding, D.M., Whitesides, G.M., Chemistry of Materials, 26[13] 2014, 3938-3947. https://doi.org/10.1021/cm5014784

[96] Morales, D., Stoute, N.A., Yu, Z., Aspnes, D.E., Dickey, M.D., Applied Physics Letters, 109[9] 2016, 091905. https://doi.org/10.1063/1.4961910

[97] Jiang, L., Sangeeth, C.S.S., Wan, A., Vilan, A., Nijhuis, C.A., Journal of Physical Chemistry C, 119[2] 2015, 960-969. https://doi.org/10.1021/jp511002b

[98] Rothemund, P., Bowers, C.M., Suo, Z., Whitesides, G.M., Chemistry of Materials, 30[1] 2018, 129-137. https://doi.org/10.1021/acs.chemmater.7b03384

[99] Kong, G.D., Kim, M., Yoon, H.J., Journal of the Electrochemical Society, 162[9] 2015, H703-H712. https://doi.org/10.1149/2.0871509jes

[100] Baghbanzadeh, M., Pieters, P.F., Yuan, L., Collison, D., Whitesides, G.M., ACS Nano, 2018, in press.

[101] Chen, X., Hu, H., Trasobares, J., Nijhuis, C.A., ACS Applied Materials and Interfaces, 11[23] 2019, 21018-21029. https://doi.org/10.1021/acsami.9b02033

[102] Baghbanzadeh, M., Belding, L., Yuan, L., Park, J., Al-Sayah, M.H., Bowers, C.M., Whitesides, G.M., Journal of the American Chemical Society, 141[22] 2019, 8969-8980. https://doi.org/10.1021/jacs.9b02891

[103] Fracasso, D., Muglali, M.I., Rohwerder, M., Terfort, A., Chiechi, R.C., Journal of Physical Chemistry C, 117[21] 2013, 11367-11376. https://doi.org/10.1021/jp401703p

[104] Wang, Z., Khalid, H., Li, B., Li, Y., Yu, X., Hu, W., Acta Chimica Sinica, 77[10] 2019, 1031-1035. https://doi.org/10.6023/A19050192

[105] Helseth, L.E., Nano Energy, 50, 2018, 266-272. https://doi.org/10.1016/j.nanoen.2018.05.047

[106] Yang, Y., Sun, N., Wen, Z., Cheng, P., Zheng, H., Shao, H., Xia, Y., Chen, C., Lan, H., Xie, X., Zhou, C., Zhong, J., Sun, X., Lee, S.T., ACS Nano, 12[2] 2018, 2027-2034. https://doi.org/10.1021/acsnano.8b00147

[107] Duan, Y., Ding, Y., Bian, J., Xu, Z., Yin, Z., Huang, Y., Polymers, 9[12] 2017, 714. https://doi.org/10.3390/polym9120714

[108] Yang, S.Y., Shih, J.F., Chang, C.C., Yang, C.R., Applied Physics A, 123[2] 2017, 128. https://doi.org/10.1007/s00339-017-0769-9

[109] Nayak, S., Li, Y., Tay, W., Zamburg, E., Singh, D., Lee, C., Koh, S.J.A., Chia, P., Thean, A.V.Y., Nano Energy, 64, 2019, 103912. https://doi.org/10.1016/j.nanoen.2019.103912

[110] Janssen, M., Werkhoven, B., Van Roij, R., RSC Advances, 6[25] 2016, 20485-20491. https://doi.org/10.1039/C5RA22814J

[111] Jeon, J., Chung, S.K., Lee, J.B., Doo, S.J., Kim, D., EPJ Applied Physics, 81[2] 2018, 20902. https://doi.org/10.1051/epjap/2018180011

[112] Chen, H., Song, Y., Han, M., Yu, B., Cheng, X., Chen, X., Chen, D., Zhang, H.,

Proceedings of the IEEE International Conference on Micro Electro Mechanical Systems, 7421865, 2016, 1252-1255.

[113] Ravindran, S.K.T., Roulet, M., Huesgen, T., Kroener, M., Woias, P., Journal of Micromechanics and Microengineering, 22[9] 2012, 094002. https://doi.org/10.1088/0960-1317/22/9/094002

[114] Zhang, N., Shen, P., Cao, Y., Guo, R.F., Jiang, Q.C., Applied Surface Science, 490, 2019, 598-603. https://doi.org/10.1016/j.apsusc.2019.06.084

[115] Eaker, C.B., Dickey, M.D., Applied Physics Reviews, 3[3] 2016, 031103. https://doi.org/10.1063/1.4959898

[116] Khan, M.R., Trlica, C., Dickey, M.D., Advanced Functional Materials, 25[5] 2015, 671-678. https://doi.org/10.1002/adfm.201403042

[117] Gough, R.C., Dang, J.H., Moorefield, M.R., Zhang, G.B., Hihara, L.H., Shiroma, W.A., Ohta, A.T., ACS Applied Materials and Interfaces, 8[1] 2016, 6-10. https://doi.org/10.1021/acsami.5b09466

[118] Xie, F., Tong, M.S., Adams, J.J., IEEE International Symposium on Antennas and Propagation and USNC-URSI Radio Science Meeting, APSURSI 2019 - Proceedings, 8888844, 2019, 459-460.

[119] Wang, M., Khan, M.R., Dickey, M.D., Adams, J.J., IEEE Antennas and Wireless Propagation Letters, 16, 2017, 79-82. https://doi.org/10.1109/LAWP.2016.2556983

[120] Cumby, B., Heikenfeld, J., Mast, D., Tabor, C., Dickey, M., IEEE Antennas and Propagation Society, AP-S International Symposium (Digest), 6904607, 2014, 553-554.

[121] Cumby, B.L., Mast, D.B., Tabor, C.E., Dickey, M.D., Heikenfeld, J., IEEE Transactions on Microwave Theory and Techniques, 63[10] 2015, 3122-3130. https://doi.org/10.1109/TMTT.2015.2470244

[122] Li, M., Anver, H.M.C.M., Zhang, Y., Tang, S.Y., Li, W., Micromachines, 10[3] 2019, 209. https://doi.org/10.3390/mi10030209

[123] Kim, D., Yoo, J.H., Lee, J.B.J.B., Choi, W., Yoo, K., IECON Industrial Electronics Conference, Proceedings, 7048829, 2014, 2340-2343.

[124] Holcomb, S., Brothers, M., Diebold, A., Thatcher, W., Mast, D., Tabor, C., Heikenfeld, J., Langmuir, 32[48] 2016, 12656-12663. https://doi.org/10.1021/acs.langmuir.6b03501

[125] Diebold, A.V., Watson, A.M., Holcomb, S., Tabor, C., Mast, D., Dickey, M.D., Heikenfeld, J., Journal of Micromechanics and Microengineering, 27[2] 2017, 025010. https://doi.org/10.1088/1361-6439/aa556a

[126] Mohseni, K., Annual IEEE Semiconductor Thermal Measurement and Management Symposium, 2005, 20-25.

[127] Jeon, J., Lee, J.B., Chung, S.K., Kim, D., 18th International Conference on Solid-State Sensors, Actuators and Microsystems, 7181305, 2015, 1834-1837.

[128] Kim, D., Lee, J.B., Journal of the Korean Physical Society, 66[2] 2015, 282-286. https://doi.org/10.3938/jkps.66.282

[129] Finkenauer, L.R., Lu, Q., Hakem, I.F., Majidi, C., Bockstaller, M.R., Langmuir, 33[38] 2017, 9703-9710. https://doi.org/10.1021/acs.langmuir.7b01322

[130] Shi, Y., Liu, L., Su, X., Wang, S., Materials Letters, 253, 2019, 9-12. https://doi.org/10.1016/j.matlet.2019.06.026

[131] Cheng, S., Rydberg, A., Hjort, K., Wu, Z., Applied Physics Letters, 94[14] 2009, 144103. https://doi.org/10.1063/1.3114381

[132] Cheng, S., Wu, Z., Hallbjörner, P., Hjort, K., Rydberg, A., IEEE Transactions on Antennas and Propagation, 57[12] 2009, 3765-3771. https://doi.org/10.1109/TAP.2009.2024560

[133] Bai, X., Su, M., Liu, Y., Wu, Y., IEEE Antennas and Wireless Propagation Letters, 17[5] 2018, 916-919. https://doi.org/10.1109/LAWP.2018.2823301

[134] Khan, M.R., Hayes, G.J., So, J.H., Lazzi, G., Dickey, M.D., Applied Physics Letters, 99[1] 2011, 013501. https://doi.org/10.1063/1.3603961

[135] Hayes, G.J., So, J.H., Qusba, A., Dickey, M.D., Lazzi, G., IEEE Transactions on Antennas and Propagation, 60[5] 2012, 2151-2156. https://doi.org/10.1109/TAP.2012.2189698

[136] Alqurashi, K.Y., Kelly, J.R., IEEE Antennas and Propagation Society International Symposium, Proceedings, 2017, 909-910.

[137] Khan, M.R., Trlica, C., So, J.H., Valeri, M., Dickey, M.D., ACS Applied Materials and Interfaces, 6[24] 2014, 22467-22473. https://doi.org/10.1021/am506496u

[138] Gurrala, P., Oren, S., Liu, P., Song, J., Dong, L., IEEE Antennas and Wireless Propagation Letters, 16, 2017, 2602-2605. https://doi.org/10.1109/LAWP.2017.2735196

[139] Wang, C., Guo, Y., Yeo, J.C., Lim, C.T., IEEE Antennas and Propagation Society

International Symposium and USNC/URSI National Radio Science Meeting, Proceedings, 8608591, 2018, 941-942.

[140] Wang, C., Yeo, J.C., Chu, H., Lim, C.T., Guo, Y.X., IEEE Antennas and Wireless Propagation Letters, 17[6] 2018, 974-977. https://doi.org/10.1109/LAWP.2018.2827404

[141] Bharambe, V., Parekh, D.P., Ladd, C., Moussa, K., Dickey, M.D., Adams, J.J., IEEE Antennas and Wireless Propagation Letters, 17[5] 2018, 739-742. https://doi.org/10.1109/LAWP.2018.2813309

[142] Xu, C., Wang, Y., Wu, J., Wang, Z., Microwave and Optical Technology Letters, 61[3] 2019, 727-733. https://doi.org/10.1002/mop.31640

[143] Song, L., Gao, W., Chui, C.O., Rahmat-Samii, Y., IEEE Transactions on Antennas and Propagation, 67[5] 2019, 2886-2895. https://doi.org/10.1109/TAP.2019.2902651

[144] Lee, M., Son, H., Lim, D., Lee, S., Lim, S., Microwave and Optical Technology Letters, 61[10] 2019, 2306-2314. https://doi.org/10.1002/mop.31898

[145] Moorefield, M.R., Gough, R.C., Morishita, A.M., Dang, J.H., Ohta, A.T., Shiroma, W.A., Electronics Letters, 52[7] 2016, 498-500. https://doi.org/10.1049/el.2015.3896

[146] Huang, G.L., Liang, J.J., Zhao, L., He, D., Sim, C.Y.D., IEEE Antennas and Wireless Propagation Letters, 18[11] 2019, 2360-2364. https://doi.org/10.1109/LAWP.2019.2932048

[147] Ramli, M.N., Soh, P.J., Rahim, S.K.A., Jamlos, M.F., Al-Hadi, A.A., Ibrahim, M.F., Lago, H., Kuster, N., IET Conference Publications, 2018, CP741.

[148] Floch, J.M., Trad, I.B., IET Conference Publications, 2017, CP732.

[149] Anwar, M.S., Bangert, A., IEEE MTT-S International Microwave Workshop Series on Advanced Materials and Processes for RF and THz Applications, 2018, 1-3.

[150] Mazlouman, S.J., Jiang, X.J., Mahanfar, A., Menon, C., Vaughan, R.G., IEEE Transactions on Antennas and Propagation, 59[12] 2011, 4406-4412. https://doi.org/10.1109/TAP.2011.2165501

[151] Aïssa, B., Nedil, M., Habib, M.A., Haddad, E., Jamroz, W., Therriault, D., Coulibaly, Y., Rosei, F., Applied Physics Letters, 103[6] 2013, 063101. https://doi.org/10.1063/1.4817861

[152] Aïssa, B., Haddad, E., Jamroz, W., Nedil, M., IEEE Antennas and Propagation Society International Symposium (Digest), 6711586, 2013, 1856-1857.

[153] Morishita, A.M., Kitamura, C.K.Y., Ohta, A.T., Shiroma, W.A., Electronics Letters, 50[1] 2014, 19-20. https://doi.org/10.1049/el.2013.2971

[154] Jeong, N.S., Koh, A., IEEE International Symposium on Antennas and Propagation and USNC-URSI Radio Science Meeting, APSURSI Proceedings, 8888557, 2019, 763-764.

[155] Gough, R.C., Morishita, A.M., Dang, J.H., Hu, W., Shiroma, W.A., Ohta, A.T., IEEE Access, 2, 2014, 874-882. https://doi.org/10.1109/ACCESS.2014.2350531

[156] Thews, J.T., Michaels, A.J., United States National Committee of URSI National Radio Science Meeting, 2017, 7878294.

[157] Zhang, G.B., Gough, R.C., Moorefield, M.R., Elassy, K.S., Ohta, A.T., Shiroma, W.A., International Journal of Antennas and Propagation, 2018, 7595363. https://doi.org/10.1155/2018/7595363

[158] Kim, D., Pierce, R.G., Henderson, R., Doo, S.J., Yoo, K., Lee, J.B., Applied Physics Letters, 105[23] 2014, 234104. https://doi.org/10.1063/1.4903882

[159] Ilyas, N., Cook, A., Tabor, C.E., Advanced Materials Interfaces, 4[15] 2017, 1700141. https://doi.org/10.1002/admi.201700141

[160] Joshipura, I.D., Ayers, H.R., Castillo, G.A., Ladd, C., Tabor, C.E., Adams, J.J., Dickey, M.D., ACS Applied Materials and Interfaces, 10[51] 2018, 44686-44695. https://doi.org/10.1021/acsami.8b13099

[161] Koo, C., Leblanc, B.E., Kelley, M., Fitzgerald, H.E., Huff, G.H., Han, A., Journal of Microelectromechanical Systems, 24[4] 2015, 1069-1076. https://doi.org/10.1109/JMEMS.2014.2381555

[162] Zandvakili, M., Honari, M.M., Sameoto, D., Mousavi, P., IEEE MTT-S International Microwave Symposium Digest, 2016, 7540271.

[163] Zandvakili, M., Honari, M.M., Mousavi, P., Sameoto, D., Advanced Materials Technologies, 2[11] 2017, 1700144. https://doi.org/10.1002/admt.201700144

[164] Bharambe, V., Adams, J.J., Joshipura, I.D., Ayers, H.R., Dickey, M.D., IEEE Antennas and Propagation Society International Symposium and USNC/URSI National Radio Science Meeting, Proceedings, 8608814, 2018, 287-288.

[165] Memon, M.U., Ling, K., Seo, Y., Lim, S., Journal of Electromagnetic Waves and Applications, 29[16] 2015, 2207-2215.

https://doi.org/10.1080/09205071.2015.1087347

[166] Memon, M.U., Lim, S., Sensors, 15[11] 2015, 28563-28573.
https://doi.org/10.3390/s151128563

[167] Eom, S., Memon, M.U., Lim, S., Sensors, 17[4] 2017, 699.
https://doi.org/10.3390/s17040699

[168] Saghati, A.P., Kordmahale, S.B., Saghati, A.P., Kameoka, J., Entesari, K., IEEE
Antennas and Propagation Society International Symposium, Proceedings,
7696131, 2016, 845-846.

[169] Thews, J., O'Donnell, A., Michaels, A.J., Proceedings - IEEE Military
Communications Conference, 2017, 812-816.

[170] Su, W., Bahr, R., Nauroze, S.A., Tentzeris, M.M., IEEE Antennas and Propagation
Society International Symposium, Proceedings, 7695943, 2016, 469-470.

[171] Singh, S., Taylor, J., Zhou, H., Pal, A., Mehta, A., Nakano, H., Howland, P., IEEE
Antennas and Propagation Society International Symposium and USNC/URSI
National Radio Science Meeting, Proceedings, 8608957, 2018, 857-858.

[172] Zhou, Y., Fang, S., Liu, H., Fu, S., International Journal of Antennas and
Propagation, 2016, 3782373. https://doi.org/10.1155/2016/3782373

[173] Zhou, Y., Fang, S., Fu, S., Wang, Z., Journal of System Simulation, 28[2] 2016,
343-347.

[174] Erdil, E., Topalli, K., Zorlu, O., Toral, T., Yildirim, E., Kulah, H., Civi, O.A., 7th
European Conference on Antennas and Propagation, 6546845, 2013, 2957-2960.

[175] Erdil, E., Topalli, K., Esmaeilzad, N.S., Zorlu, O., Kulah, H., Civi, O.A., 8th
European Conference on Antennas and Propagation, 6901708, 2014, 124-127.

[176] Erdil, E., Topalli, K., Esmaeilzad, N.S., Zorlu, O., Kulah, H., Civi, O.A., IEEE
Transactions on Antennas and Propagation, 63[3] 2015, 1163-1167.
https://doi.org/10.1109/TAP.2014.2387424

[177] Zhang, Y., Lin, S., Sun, Z., Li, Y., Chen, Z., Yang, C., Zhang, H., Denisov, A.,
IEEE International Symposium on Antennas and Propagation and USNC-URSI
Radio Science Meeting, APSURSI 2019 - Proceedings, 8889303, 2019, 1875-
1876.

[178] Li, M., Behdad, N., IEEE Transactions on Antennas and Propagation, 60[6] 2012,
2748-2759. https://doi.org/10.1109/TAP.2012.2194645

[179] Wang, M., Kilgore, I.M., Steer, M.B., Adams, J.J., IEEE Antennas and Wireless

Propagation Letters, 17[2] 2018, 279-282.
https://doi.org/10.1109/LAWP.2017.2786078

[180] Moghadas, H., Zandvakili, M., Sameoto, D., Mousavi, P., IEEE Antennas and
Wireless Propagation Letters, 16, 2017, 1337-1340.
https://doi.org/10.1109/LAWP.2016.2633964

[181] Bharambe, V., Parekh, D.P., Ladd, C., Moussa, K., Dickey, M.D., Adams, J.J.,
Additive Manufacturing, 18, 2017, 221-227.
https://doi.org/10.1016/j.addma.2017.10.012

[182] Lin, Y., Gordon, O., Khan, M.R., Vasquez, N., Genzer, J., Dickey, M.D., Lab on a
Chip, 17[18] 2017, 3043-3050. https://doi.org/10.1039/C7LC00426E

[183] Smith, L.M., Pan, H.K., Bradford, R.L., Frank, G.J., Hartl, D., Baur, J., Huff,
G.H., IEEE Antennas and Propagation Society International Symposium,
Proceedings, 7696132, 2016, 847-848.

[184.] Vahabisani, N., Khan, S., Daneshmand, M., IEEE Antennas and Wireless
Propagation Letters, 16, 2017, 1788-1791.
https://doi.org/10.1109/LAWP.2017.2782925

[185] Liu, P., Yang, S., Wang, X., Yang, M., Song, J., Dong, L., IEEE Antennas and
Wireless Propagation Letters, 16, 2017, 66-69.
https://doi.org/10.1109/LAWP.2016.2555941

[186] Kim, D., Doo, S.J., Won, H.S., Lee, W., Jeon, J., Chung, S.K., Lee, G.Y., Oh, S.,
Lee, J.B., EPJ Applied Physics, 78[1] 2017, 11101.
https://doi.org/10.1051/epjap/2017160346

[187] Zhang, G.B., Gough, R.C., Moorefield, M.R., Cho, K.J., Ohta, A.T., Shiroma,
W.A., IEEE Antennas and Wireless Propagation Letters, 17[1] 2018, 50-53.
https://doi.org/10.1109/LAWP.2017.2773076

[188] Sarabia, K.J., Yamada, S.S., Moorefield, M.R., Combs, A.W., Ohta, A.T.,
Shiroma, W.A., International Journal of Antennas and Propagation, 2018,
1248459. https://doi.org/10.1155/2018/1248459

[189] Sarabia, K.J., Ohta, A.T., Shiroma, W.A., Electronics Letters, 55[19] 2019, 1032-
1034. https://doi.org/10.1049/el.2019.0765

[190] Huff, G.H., Pan, H., Hartl, D.J., Frank, G.J., Bradford, R.L., Baur, J.W., IEEE
Transactions on Antennas and Propagation, 65[5] 2017, 2282-2288.
https://doi.org/10.1109/TAP.2017.2679075

[191] Hartl, D.J., Frank, G.J., Huff, G.H., Baur, J.W., Smart Materials and Structures, 26[2] 2017, 025001. https://doi.org/10.1088/1361-665X/aa5142

[192] Hartl, D.J., Frank, G.J., Malak, R.J., Baur, J.W., Smart Materials and Structures, 26[2] 2017, 025002. https://doi.org/10.1088/1361-665X/aa513d

[193] Baur, J.W., Gibson, T., Rapking, D., Murphy, S., Frank, G.J., Bradford, R., Huff, G., Hartl, D.J., Phillips, D., International SAMPE Technical Conference, 2017, 2477-2486.

[194] Baur, J.W., Hartl, D.J., Frank, G.J., Huff, G., Slinker, K.A., Kondash, C., Kennedy, W.J., Ehlert, G.J., Conference Proceedings of the Society for Experimental Mechanics Series, 6, 2018, 215-221. https://doi.org/10.1007/978-3-319-63408-1_21

[195] Su, W., Nauroze, S.A., Ryan, B., Tentzeris, M.M., IEEE MTT-S International Microwave Symposium Digest, 8058933, 2017, 1579-1582.

[196] Su, W., Bahr, R., Nauroze, S.A., Tentzeris, M.M., IEEE Antennas and Propagation Society International Symposium, Proceedings, 2017, 1245-1246.

[197] Lee, D., Seo, Y., Lim, S., Microwave and Optical Technology Letters, 58[3] 2016, 668-672. https://doi.org/10.1002/mop.29647

[198] Ghosh, S., Hussain Shah, S.I., Lim, S., IEEE Antennas and Propagation Society International Symposium and USNC/URSI National Radio Science Meeting, Proceedings, 8608246, 2018, 2045-2046.

[199] Ghosh, S., Lim, S., IEEE Transactions on Antennas and Propagation, 66[9] 2018, 4953-4957. https://doi.org/10.1109/TAP.2018.2851455

[200] Ghosh, S., Lim, S., IEEE Transactions on Microwave Theory and Techniques, 66[8] 2018, 3857-3865. https://doi.org/10.1109/TMTT.2018.2829195

[201] Li, R., Guo, Y.X., Chen, W., Li, Y., Thean, A.V.Y., International Journal of RF and Microwave Computer-Aided Engineering, 28[5] 2018, e21265. https://doi.org/10.1002/mmce.21265

[202] Psychogiou, D., Sadasivan, K., IEEE Microwave and Wireless Components Letters, 29[12] 2019, 763-766. https://doi.org/10.1109/LMWC.2019.2950540

[203] Alqurashi, K.Y., Crean, C., Filgueiras, H.R.D., Da Costa, I.F., Arismar Cerqueira, S., Xiao, P., Chen, Z., Wong, H., Kelly, J.R., Proceedings of the 8th IEEE-APS Topical Conference on Antennas and Propagation in Wireless Communications, 8503677, 2018, 814-815.

[204] Gheethan, A.A., Jo, M.C., Guldiken, R., Mumcu, G., IEEE Antennas and Wireless Propagation Letters, 12, 2013, 1638-1641. https://doi.org/10.1109/LAWP.2013.2294153

[205] Gheethan, A., Guldiken, R., Mumcu, G., IEEE Antennas and Propagation Society, International Symposium (Digest), 6710765, 2013, 208-209.

[206] Gheethan, A.A., Dey, A., Mumcu, G., United States National Committee of URSI National Radio Science Meeting, 2014, 6927940.

[207] Xie, F., Tong, M.S., Progress in Electromagnetics Research Symposium, 8598010, 2018, 2314-2317.

[208] Wang, M., Trlica, C., Khan, M.R., Dickey, M.D., Adams, J.J., Journal of Applied Physics, 117[19] 2015, 194901. https://doi.org/10.1063/1.4919605

[209] He, Z., Hua, Z., Hongmei, L., Fan, S., Shu, L., Denisov, A., International Symposium on Antennas and Propagation, 2017, 1-2.

[210] Zheng, P., Zhang, B., Duan, J., Wang, W., Tian, Y., Proceedings of IEEE 3rd Advanced Information Technology, Electronic and Automation Control Conference, 8577727, 2018, 1244-1247.

[211] Shah, S.I.H., Lim, S., Sensors, 18[9] 2018, 2935. https://doi.org/10.3390/s18092935

[212] Morishita, A.M., Kitamura, C.K.Y., Ohta, A.T., Shiroma, W.A., IEEE Antennas and Wireless Propagation Letters, 12, 2013, 1388-1391. https://doi.org/10.1109/LAWP.2013.2286544

[213] Kitamura, C.K.Y., Morishita, A.M., Chun, T.F., Tonaki, W.G., Ohta, A.T., Shiroma, W.A., IEEE MTT-S International Microwave Symposium Digest, 2013, 6697779.

[214] Zheng, P.S., Chen, J., Yao, P., Cui, J.L., Zhang, B.Z., Duan, J.P., Wang, W.J., Acta Electronica Sinica, 46[9] 2018, 2276-2282.

[215] Lee, M., Lim, S., Sensors, 18[10] 2018, 3176. https://doi.org/10.3390/s18103176

[216] Cosker, M., Lizzi, L., Ferrero, F., Staraj, R., Ribero, J.M., IEEE Antennas and Wireless Propagation Letters, 16, 2017, 971-974. https://doi.org/10.1109/LAWP.2016.2615568

[217] Ha, A., Kim, K., Electronics Letters, 52[2] 2016, 100-102. https://doi.org/10.1049/el.2015.3009

[218] Chen, Z., Wong, H., Kelly, J., IEEE Transactions on Antennas and Propagation,

67[5] 2019, 3427-3432. https://doi.org/10.1109/TAP.2019.2901132

[219] Jackson, N., Buckley, J., Clarke, C., Stam, F., Microsystem Technologies, 25[8] 2019, 3175-3184. https://doi.org/10.1007/s00542-018-4234-2

[220] Matsubara, K., Ota, H., Proceedings of the IEEE International Conference on Micro Electro Mechanical Systems, 8870779, 2019, 296-298.

[221] Ha, A., Chae, M.H., Kim, K., IEEE Antennas and Wireless Propagation Letters, 18[4] 2019, 571-575. https://doi.org/10.1109/LAWP.2019.2894397

[222] Zhu, J., Fox, J.J., Yi, N., Cheng, H., ACS Applied Materials and Interfaces, 11[9] 2019, 8867-8877. https://doi.org/10.1021/acsami.8b22021

[222] Alqurashi, K.Y., Crean, C., Kelly, J.R., Brown, T.W.C., Khalily, M., 13th European Conference on Antennas and Propagation, 2019, 8740214.

[224] Zhou, X., He, Y., Zeng, J., Smart Materials and Structures, 28[2] 2019, 025019. https://doi.org/10.1088/1361-665X/aaf842

[225] Moorefield, M.R., Ohta, A.T., Shiroma, W.A., Asia-Pacific Microwave Conference Proceedings, 8617330, 2019, 1250-1252.

[226] Wang, F., Arslan, T., 2017 IEEE MTT-S International Microwave Workshop Series on Advanced Materials and Processes for RF and THz Applications, 2018, 1-3.

[227] Dey, A., Guldiken, R., Mumcu, G., IEEE Transactions on Antennas and Propagation, 64[6] 2016, 2572-2576. https://doi.org/10.1109/TAP.2016.2551358

[228] Dey, A., Guldiken, R., Mumcu, G., IEEE Antennas and Propagation Society, International Symposium (Digest), 6710857, 2013, 392-393.

[229] Wang, M., Khan, M.R., Trlica, C., Dickey, M.D., Adams, J.J., IEEE Antennas and Propagation Society, AP-S International Symposium (Digest), 7305500, 2015, 2223-2224.

[230] Saghati, A.P., Batra, J.S., Kameoka, J., Entesari, K., IEEE Antennas and Wireless Propagation Letters, 15, 2016, 122-125. https://doi.org/10.1109/LAWP.2015.2432773

[231] Saghati, A.P., Batra, J.S., Kameoka, J., Entesari, K., IEEE Transactions on Antennas and Propagation, 63[9] 2015, 3798-3807. https://doi.org/10.1109/TAP.2015.2447002

[232] Dang, J.H., Gough, R.C., Morishita, A.M., Ohta, A.T., Shiroma, W.A., Electronics Letters, 51[21] 2015, 1630-1632. https://doi.org/10.1049/el.2015.2782

[233] Hu, A., Shaikh, K., Liu, C., Proceedings of IEEE Sensors, 4388525, 2007, 815-817.

[234] Traille, A., Bouaziz, S., Pinon, S., Pons, P., Aubert, H., Boukabache, A., Tentzeris, M., Conference Proceedings, 41st European Microwave Conference, 6101891, 2011, 45-48.

[235] Bouaziz, S., Pons, P., Aubert, H., Traille, A., Tentzeris, M.M., Proceeding of the 13th International Conference on Microwave and Radio Frequency Heating, 2011, 153-156.

[236] Cheng, S., Wu, Z., Advanced Functional Materials, 21[12] 2011, 2282-2290. https://doi.org/10.1002/adfm.201002508

[237] Shi, X., Cheng, C.H., Chao, C., Wang, L., Zheng, Y., 7th IEEE International Conference on Nano/Micro Engineered and Molecular Systems, 6196822, 2012, 483-486.

[238] Wong, R.D.P., Posner, J.D., Santos, V.J., Sensors and Actuators A, 179, 2012, 62-69. https://doi.org/10.1016/j.sna.2012.03.023

[239] Shi, X., Cheng, C.H., 8th Annual IEEE International Conference on Nano/Micro Engineered and Molecular Systems, 6559886, 2013, 978-981.

[240] Fernandez, E., Carandang, A., Orillo, J.W., Quezada, L.R., Valenzuela, I., Yee, V., 7th HNICEM, 6th International Symposium on Computational Intelligence and Intelligent Informatics and 10th ERDT Conference, 2014, 7016229.

[241] Fernandez, E.O., Valenzuela, I., Orillo, J.W., Jurnal Teknologi, 78[5-9] 2016, 79-84. https://doi.org/10.11113/jt.v78.8720

[242] Shi, X., Cheng, C.H., Zheng, Y., Wai, P.K.A., Journal of Micromechanics and Microengineering, 26[10] 2016, 105020. https://doi.org/10.1088/0960-1317/26/10/105020

[243] Yan, H.L., Chen, Y.Q., Deng, Y.Q., Zhang, L.L., Hong, X., Lau, W.M., Mei, J., Hui, D., Yan, H., Liu, Y., Applied Physics Letters, 109[8] 2016, 083502. https://doi.org/10.1063/1.4961493

[244] Lin, Y., Cooper, C., Wang, M., Adams, J.J., Genzer, J., Dickey, M.D., Small, 11[48] 2015, 6397-6403. https://doi.org/10.1002/smll.201502692

[245] Vorobyov, A., Henemann, C., Dallemagne, P., 10th European Conference on Antennas and Propagation, 2016, 7481418.

[246] Hellebrekers, T., Ozutemiz, K.B., Yin, J., Majidi, C., IEEE International

Conference on Intelligent Robots and Systems, 8593944, 2018, 5924-5929.

[247] Rahimi, R., Yu, W., Ochoa, M., Ziaie, B., Lab on a Chip, 17[9] 2017, 1585-1593. https://doi.org/10.1039/C7LC00074J

[248] Zatopa, A., Walker, S., Menguc, Y., Soft Robotics, 5[3] 2018, 258-271. https://doi.org/10.1089/soro.2017.0019

[249] Mitraka, E., Kergoat, L., Khan, Z.U., Fabiano, S., Douhéret, O., Leclère, P., Nilsson, M., Andersson Ersman, P., Gustafsson, G., Lazzaroni, R., Berggren, M., Crispin, X., Journal of Materials Chemistry C, 3[29] 2015, 7604-7611. https://doi.org/10.1039/C5TC00753D

[250] Khondoker, M.A.H., Ostashek, A., Sameoto, D., Advanced Engineering Materials, 21[7] 2019, 1900060. https://doi.org/10.1002/adem.201900060

[251] Guo, R., Sun, X., Yuan, B., Wang, H., Liu, J., Advanced Science, 2019, 1901478. https://doi.org/10.1002/advs.201901478

[252] Teng, L., Ye, S., Handschuh-Wang, S., Zhou, X., Gan, T., Zhou, X., Advanced Functional Materials, 29[11] 2019, 1808739. https://doi.org/10.1002/adfm.201808739

[253] Deshpande, R.D., Li, J., Cheng, Y.T., Verbrugge, M.W., Journal of the Electrochemical Society, 158[8] 2011, A845-A849. https://doi.org/10.1149/1.3591094

[254] Parekh, D.P., Ladd, C., Panich, L., Moussa, K., Dickey, M.D., Lab on a Chip, 16[10] 2016, 1812-1820. https://doi.org/10.1039/C6LC00198J

[255] Tavakoli, M., Malakooti, M.H., Paisana, H., Ohm, Y., Green Marques, D., Alhais Lopes, P., Piedade, A.P., de Almeida, A.T., Majidi, C., Advanced Materials, 30[29] 2018, 1801852. https://doi.org/10.1002/adma.201801852

[256] Blaiszik, B.J., Jones, A.R., Sottos, N.R., White, S.R., Journal of Microencapsulation, 31[4] 2014, 350-354. https://doi.org/10.3109/02652048.2013.858790

[257] Jeong, Y.R., Kim, J., Xie, Z., Xue, Y., Won, S.M., Lee, G., Jin, S.W., Hong, S.Y., Feng, X., Huang, Y., Rogers, J.A., Ha, J.S., NPG Asia Materials, 9[10] 2017, e443. https://doi.org/10.1038/am.2017.189

[258] Rich, S., Jang, S.H., Park, Y.L., Majidi, C., Advanced Materials Technologies, 2[12] 2017, 1700179. https://doi.org/10.1002/admt.201700179

[259] Yu, Z., Shang, J., Niu, X., Liu, Y., Liu, G., Dhanapal, P., Zheng, Y., Yang, H.,

Wu, Y., Zhou, Y., Wang, Y., Tang, D., Li, R.W., Advanced Electronic Materials, 4[9] 2018, 1800137. https://doi.org/10.1002/aelm.201800137

[260] Jeong, S.H., Hjort, K., Wu, Z., Sensors, 14[9] 2014, 16311-16321. https://doi.org/10.3390/s140916311

[261] Li, B., Gao, Y., Fontecchio, A., Visell, Y., Smart Materials and Structures, 25[7] 2016, 075009. https://doi.org/10.1088/0964-1726/25/7/075009

[262] Wang, L., Liu, J., RSC Advances, 5[71] 2015, 57686-57691. https://doi.org/10.1039/C5RA10295B

[263] Zhang, Y., Tang, S.Y., Zhao, Q., Yun, G., Yuan, D., Li, W., Applied Physics Letters, 114[15] 2019, 154101. https://doi.org/10.1063/1.5086376

[264] Lazarus, N., Bedair, S.S., Kierzewski, I.M., ACS Applied Materials and Interfaces, 9[2] 2017, 1178-1182. https://doi.org/10.1021/acsami.6b13088

[265] Kubo, M., Li, X., Kim, C., Hashimoto, M., Wiley, B.J., Ham, D., Whitesides, G.M., Electronic Components and Technology Conference Proceedings, 5898722, 2011, 1582-1587.

[266] Kubo, M., Li, X., Kim, C., Hashimoto, M., Wiley, B.J., Ham, D., Whitesides, G.M., Advanced Materials, 22[25] 2010, 2749-2752. https://doi.org/10.1002/adma.200904201

[267] Guo, R., Tang, J., Dong, S., Lin, J., Wang, H., Liu, J., Rao, W., Advanced Materials Technologies, 3[12] 2018, 1800265. https://doi.org/10.1002/admt.201800265

[268] Mineart, K.P., Lin, Y., Desai, S.C., Krishnan, A.S., Spontak, R.J., Dickey, M.D., Soft Matter, 9[32] 2013, 7695-7700. https://doi.org/10.1039/c3sm51136g

[269] Lopes, P.A., Paisana, H., De Almeida, A.T., Majidi, C., Tavakoli, M., ACS Applied Materials and Interfaces, 10[45] 2018, 38760-38768. https://doi.org/10.1021/acsami.8b13257

[270] Kim, D., Yoo, J.H., Lee, Y., Choi, W., Yoo, K., Lee, J.B.J., Proceedings of the IEEE International Conference on Micro Electro Mechanical Systems, 6765804, 2014, 967-970.

[271] Guo, R., Sun, X., Yao, S., Duan, M., Wang, H., Liu, J., Deng, Z., Advanced Materials Technologies, 4[8] 2019, 1900183. https://doi.org/10.1002/admt.201900183

[272] Marques, D.G., Lopes, P.A., De Almeida, A.T., Majidi, C., Tavakoli, M., Lab on a

Chip, 19[5] 2019, 897-906. https://doi.org/10.1039/C8LC01093E

[273] Khondoker, M.A.H., Sameoto, D., Smart Materials and Structures, 25[9] 2016, 093001. https://doi.org/10.1088/0964-1726/25/9/093001

[274] Secor, E.B., Cook, A.B., Tabor, C.E., Hersam, M.C., Advanced Electronic Materials, 4[1] 2018, 1700483. https://doi.org/10.1002/aelm.201700483

[275] Ozutemiz, K.B., Wissman, J., Ozdoganlar, O.B., Majidi, C., Advanced Materials Interfaces, 5[10] 2018, 1701596. https://doi.org/10.1002/admi.201701596

[276] Fisher, D.J., Materials Research Foundations, 67, 2020, 1-154. https://doi.org/10.21741/9781644900637

[277] Swensen, J.P., Odhner, L.U., Araki, B., Dollar, A.M., Proceedings - IEEE International Conference on Robotics and Automation, 7139297, 2015, 988-995.

[278] Swensen, J.P., Odhner, L.U., Araki, B., Dollar, A.M., Journal of Mechanisms and Robotics, 7[2] 2015, 021004. https://doi.org/10.1115/1.4029435

[279] Gannarapu, A., Gozen, B.A., Extreme Mechanics Letters, 33, 2019, 100554. https://doi.org/10.1016/j.eml.2019.100554

[280] Park, Y.G., An, H.S., Kim, J.Y., Park, J.U., Science Advances, 5[6] 2019, 2844.

[281] Moon, Y.G., Koo, J.B., Park, N.M., Oh, J.Y., Na, B.S., Lee, S.S., Ahn, S.D., Park, C.W., IEEE Transactions on Electron Devices, 64[12] 2017, 5157-5162. https://doi.org/10.1109/TED.2017.2758784

[282] Qusba, A., Ramrakhyani, A.K., So, J.H., Hayes, G.J., Dickey, M.D., Lazzi, G., IEEE Sensors Journal, 14[4] 2014, 1074-1080. https://doi.org/10.1109/JSEN.2013.2293096

[283] Latif, T., Gong, F., Dickey, M., Sichitiu, M., Bozkurt, A., Proceedings of the Annual International Conference of the IEEE Engineering in Medicine and Biology Society, 7591152, 2016, 2141-2144.

[284] Khan, M.R., Hayes, G.J., Zhang, S., Dickey, M.D., Lazzi, G., IEEE Microwave and Wireless Components Letters, 22[11] 2012, 577-579. https://doi.org/10.1109/LMWC.2012.2223754

[285] Park, E., Lee, M., Lim, S., Sensors, 19[5] 2019, 1081. https://doi.org/10.3390/s19051081

[286] Ling, K., Kim, H.K., Yoo, M., Lim, S., Sensors, 15[11] 2015, 28154-28165. https://doi.org/10.3390/s151128154

[287] Kim, H.K., Lee, D., Lim, S., Scientific Reports, 6, 2016, 31823.
https://doi.org/10.1038/srep31823

[288] Kim, H.K., Lim, S., 2016 IEEE Antennas and Propagation Society International
Symposium, Proceedings, 7695956, 2016, 495-496.

[289] Finkenauer, L.R., Majidi, C., Proceedings of SPIE, 2014, 90563I.

[290] Elassy, K.S., Akau, T.K., Shiroma, W.A., Seo, S., Ohta, A.T., Applied Sciences,
9[8] 2019, 1565. https://doi.org/10.3390/app9081565

[291] Liu, D., Su, L., Liao, J., Reeja-Jayan, B., Majidi, C., Advanced Energy Materials,
1902798, 2019. https://doi.org/10.1002/aenm.201902798

Keyword Index

alumina, 3, 19

aluminium, 2, 19, 96

amphiphilic, 27

battery, 107

beamwidth, 81

body, 18, 21, 75, 87, 93

buckling, 49

buffer, 18, 30

capacitance, 41, 45, 50

ceramic, 8

channel, 5, 10-14, 21, 24, 30, 53-54, 58,
62-64, 66-68, 70-74, 78, 81-83, 99-
100, 103

charge transport, 35, 37, 39-40, 44

chemisorbed, 37

colloidal, 10

compliant, 30, 92

coolant, 10,-13, 20, 24, 30

copper, 17, 25, 30, 51-53, 62, 67, 72, 74,
76, 82, 87, 97-100

curable, 105

degradation, 21, 43, 94-95, 100

dielectrophoretic, 7

diffraction, 55

dipole, 37, 40, 46-47, 55, 62, 68, 71, 73-
76, 78-79, 81-82, 85

director, 71, 79

droplets, 5-6, 10, 14-16, 19, 21, 33, 51,
57, 59-71, 88, 92, 95

dual-scale structure, 4

EGaIn, iii, 3, 6, 8-9, 14-15, 21-23, 27-28,
37-42, 44-47, 52, 54, 56-57, 60, 62-63,
65, 67-68, 70-71, 76-79, 82, 85-88, 91-
92, 94-101, 103-104, 107

elastic modulus, 2-3, 27

elastomer, 4-8, 20, 30, 39, 48-51, 62, 64,
68, 73, 79, 83-88, 92, 97, 104, 106

electrical resistivity, 10, 95

electrochemical, 1, 17, 19, 34, 42, 52, 55,
57-58, 78, 87, 89, 90, 107

electrode, 7, 13, 15, 17-18, 30, 35, 37,
39, 41-50, 58, 87, 89, 98, 104

electrolyte, 44, 52-54, 58, 66, 68, 82

electrophoresis, 29

etching, 6, 16, 45, 58

ethanol, 19, 22, 60

eutectic, 6, 9, 15, 19-20, 26, 35-37, 43,
47, 59, 61-62, 64-65, 73, 76, 81, 96,
132

flexible, 6, 9, 20, 29-30, 50, 52, 55, 60,
62, 64, 70, 72, 76, 79-80, 83-85, 87-88,
91-93, 95, 99, 103, 106

foil, 4

folding, 49, 76

functional groups, 17, 40, 46-47

galinstan, 7, 10-21, 24, 26-27, 29-30, 32-
33, 49, 51, 64-66, 74-75, 77, 80, 83-84,
87, 94, 96, 106

gallium, 1-3, 5-6, 8-11, 14-15, 17, 19-27,
32, 34-35, 43, 47, 49-50, 54, 58-59, 61-
65-67, 72-74, 76, 80-81, 83, 87-89, 92,
94-98, 107

gallium oxide, 14, 17, 21-22, 27, 54, 74,
97

gold, 6-7, 16-17, 19, 26, 37, 38, 41, 45,
47, 98-100

grating, 55

humidity, 8, 38-39, 43

hydrogel, 80, 107

indentation, 27, 87

indium, 2-3, 5-6, 8-10, 15, 20, 22, 24,
26-27, 35, 43, 45, 47, 49, 59, 61-65,
73, 76, 80-81, 87-88, 92, 96, 98

injection, 7-8, 17, 37, 39-40, 44, 58, 68-
69, 72, 80, 88

interconnect, 75, 83
isothermalization, 11

Joule heating, 44
junction, 11, 29, 37, 43, 44, 46

laser, 7, 26, 27, 30, 51, 57, 95-96, 103
leakage, 20, 45-46, 48
Lewis acid, 40

mask, 27, 51, 106
melting, 1-2, 10, 22, 26, 37, 89, 97
mercury, 1-2, 8-9, 13, 18, 20
microfluidic, 7, 9, 13, 17, 21, 27, 52-53, 58, 61-63, 67-68, 70-71, 77-85, 87, 93-94, 100-101, 103
microspring, 81
microwelding, 29
molybdenum, 30
monolayer, 26, 36-47, 60
motors, 29

nanogenerator, 49
nanorobots, 29
nanotube, 65, 76
nanowire, 29, 106
needle, 8, 73, 88, 106

parasitic, 63, 70-71, 79
patch, 62-65, 68, 81, 84, 93, 106
phased-array, 77
phosphonic acid, 28, 29
plasmonic, 22, 76
platinum, 7, 19, 29
polarizer, 55, 58, 80
polydimethylsiloxane, 5-7, 10, 13, 15-16, 21, 27, 43, 45, 58-59, 61-65, 67, 69, 71, 73-74, 77-85, 94-95, 101, 103-104
polyimide, 74, 105-106
potassium, 6, 30
pressure, 1, 3, 6, 9, 11-12, 20-21, 44, 61-62, 67-68, 72, 74, 79, 80-81, 83, 86-88, 90-91, 97, 100, 102, 105-106

pumping, 12, 52, 59, 78

reflector, 71, 81
rheometer, 3, 21
roughening, 45

short-circuit, 6, 23, 49-50, 83
silver, 9, 16, 18-19, 22, 29-30, 39-40, 41, 44-46, 91, 95-96, 98-99
skin, 1, 3, 13, 16, 26, 36, 45, 48, 53, 61, 83, 87, 90-92, 95-98
slug, 58, 63, 66-67, 74
slurry, 8, 107
sodium, 2, 16, 18, 30, 57, 64, 66, 74
solidification, 10, 31, 132
sonication, 6, 19, 22-23, 27
steerable, 77
stencil, 51, 106
stiction, 21
stretchable, 1, 6, 8-9, 16, 20, 22, 43, 49, 65, 73, 79, 81, 84-87, 93-98, 107
stub, 67, 70, 80, 100-101
substrate, 4-6, 9, 12, 15, 24-26, 28, 35, 48, 51-52, 58, 60, 62, 65, 67-70, 74, 76-77, 81-82, 84-85, 88, 94, 97, 106
super-hydrophobic, 4-5, 16
super-lyophobic, 4, 13-14
surface stress, 2-3, 21
surface tension, 1, 10, 12, 14-16, 27, 46, 54, 58-59, 64, 66, 74, 78, 82, 88, 94
surfusion, 3

template-stripped, 37, 39, 45, 47
thermal conductivity, 9, 11-12, 20, 25, 30
thermal resistance, 10, 12
thermofluidic, 97
thiol, 37, 39, 47, 60
tin, 4, 7, 9, 10, 15, 20, 25-26, 45, 80, 87, 89
toxicity, 1, 9, 13, 19
tribo-electric, 48-49
tunnelling, 35, 37-42, 44-47

twisting, 49, 77, 87-88

undercooling, 92

vibration, 50

waveguide, 67-71, 73, 79, 82, 103

wax, 8, 72
wettability, 4-5, 14, 16, 23, 30, 88, 94
wetting, 3-6, 12-15, 21, 23, 26, 39, 41,
   52, 59-60, 67, 74, 92

Young-Laplace equation, 50

About the author

Dr Fisher has wide knowledge and experience of the fields of engineering, metallurgy and solid-state physics, beginning with work at Rolls-Royce Aero Engines on turbine-blade research, related to the Concord supersonic passenger-aircraft project, which led to a BSc degree (1971) from the University of Wales. This was followed by theoretical and experimental work on the directional solidification of eutectic alloys having the ultimate aim of developing composite turbine blades. This work led to a doctoral degree (1978) from the Swiss Federal Institute of Technology (Lausanne). He then acted for many years as an editor of various academic journals, in particular *Defect and Diffusion Forum*. In recent years he has specialised in writing monographs which introduce readers to the most rapidly developing ideas in the fields of engineering, metallurgy and solid-state physics. His latest paper will appear shortly in *International Materials Reviews*, and he is co-author of the widely-cited student textbook, *Fundamentals of Solidification*.

www.ingramcontent.com/pod-product-compliance
Lightning Source LLC
Chambersburg PA
CBHW071705210326
41597CB00017B/2340